Topics in Recreational Mathematics 5/2015

Editor-in-chief

Charles Ashbacher
5530 Kacena Ave
Marion, IA 52302 USA

cashbacher@yahoo.com

Assistant Editors

Rachel Pollari

Jennifer Corrigan

Artwork

Caytie Ribble

Problems

Lamarr Widmer

ISBN: 978-1519115676

CONTENTS

Note From the Editor

Welcome to the fifth in our series of "Topics in Recreational Mathematics" books. Like the previous books, there are two math cartoons by Caytie Ribble and I have also included an old and a new math limerick. The second is a continuation of the first and I hope you enjoy the second limerick that was developed as a consequence of a challenge.

This issue also includes a complete index to the material that appeared in **Recreational Mathematics Magazine (RMM)** that was edited and published by Joseph S. Madachy. The index is in three separate sections, the first two were prepared by Joe and were published in **RMM**. The last section is an index to the last two issues and was prepared by me. I consider it important that this material be preserved and there was not enough to justify a book, so the decision was made to publish it here.

It is my hope that you, the reader, enjoy this book and find something in it that will entertain you and spur you to do more work in mathematics. If you are an educator, perhaps you will find something that you can use in your classes. Problems and solutions to problems are always welcome, Lamarr Widmer has agreed to keep that section of the books going. Contact him at his address given in the problem section if you wish to contribute. As always, I welcome feedback and comments via my email address.

Charles Ashbacher

cashbacher@yahoo.com

Mathematical Limerick

Jen Corrigan

The first limerick has been around for some time with the writer unknown. Recently, it was republished on the Facebook page "Math Believe it or Not" (https://www.facebook.com/groups/maths.believe.it.or.not/) followed by a statement questioning what happened next in the student's life. The second limerick was composed by Jen Corrigan as a continuation of the story.

A formidable student at Trinity
Solved the square root of infinity
It gave him such fidgets
To count up the digits
He chucked math and took up divinity - Anonymous

Did the student continue at Trinity
After he took up divinity?
There was really no way
In good conscience he'd stay
He said, "See ya," and left the vicinity.

Mathematical Cartoons

Caytie Ribble

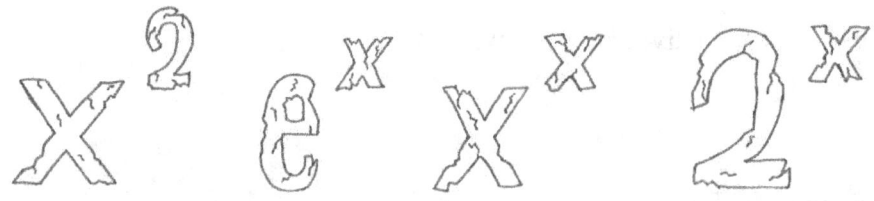

FORCE OF
GRAVITY

32.2 Ft/s^2

9.8 m/s^2

HERE COMES THE SUN. GIMME SHELTER?

Paul M. Sommers

Department of Economics

Middlebury College

Middlebury, Vermont 05753

psommers@middlebury.edu

Abstract

Based on daily ultraviolet (UV) index forecasts over three years (2006 through 2008) in fifty-seven cities (at least one for each state), this paper examines the relationship between melanoma skin cancer incidence rates and sun exposure. There is no evidence of a positive association between the two variables. When the sample is divided into two groups, cities above the 37th parallel (associated with less sun exposure) and those below, there is surprisingly no statistically discernible difference between the two average skin cancer rates. And, when the sample is divided into four census regions, skin cancer incidence rates are, on average, highest not in the South or West, but in the Northeast.

Fear of skin cancer is one of the main

reasons for the hysteria over sun exposure.

— Dr. Michael F. Holick [1, p. 29]

"Sun-phobes" believe that exposure to the sun increases one's chances of getting skin cancer. People who live in southern latitudes are therefore probably at greater risk than those who live in northern latitudes. After all, they are exposed to more of the sun's ultraviolet or UV radiation than are their neighbors to the north.[1] One would therefore expect to find a strong direct association between a city's skin cancer incidence rate and its sun exposure over the course of a year. If, for example, one were to divide the U.S. into four census regions, one would expect to find the most UV radiation and hence the highest skin cancer incidence rates in the South. This paper examines these beliefs against recent historical data for fifty-seven U.S. cities.

Daily UV Index forecasts over three years (2006 through 2008) were collected on fifty-seven cities (one for each state and the District of Columbia, with two observations each for the states of California [Los Angeles and San Francisco], Texas [Dallas and Houston], New York [Buffalo and New York City], and Pennsylvania [Pittsburgh and Philadelphia] and three observations for Florida [Jacksonville, Miami, and Tampa Bay]). The UV Index ranges from 1 to 15 and its five categories are "Low" (UV Index of 0, 1, or 2), "Moderate" (3, 4, or 5), "High" (6, 7), "Very High" (8, 9, 10), and "Extreme" (11 or higher). The annual (2006 through 2008) time series of UV Index forecasts are from the National Weather Service website at http://www.cpc.ncep.noaa.gov/products/stratosphere/uv_index/index.html. There are three types of skin cancer: the two mildest forms are basal and squamous; the most dangerous form of skin cancer is melanoma. Metropolitan statistical area (MSA) level cancer data are available from the Centers for Disease Control and Prevention (CDC) at http://wonder.cdc.gov/cancer.html.[2] The average (2006 through 2008) skin cancer (excluding basal and squamous) incidence rates (age-adjusted rate per 100,000 people) were collected for each of the MSAs to which the aforementioned fifty-seven cities belong.[3] A latitude locator for each city is from http://www.travelmath.com/cities/ .

Table 1 shows the latitude, three-year (2006 through 2008) average skin cancer incidence rate, and three-year (2006 through 2008) daily UV Index forecast for each of fifty-seven U.S. cities. The scatterplot in Figure 1 shows the relationship between skin cancer incidence rate and the percentage of days (over the three-year period 2006 through 2008) that the daily UV Index forecast was either "High," "Very High" or "Extreme" (hereafter the *UV Intensity*). The two obvious outliers are Minneapolis, Minnesota with close to a zero skin cancer incidence rate and Anchorage, Alaska with no days over the three-year period that the UV Index forecast was ever above "Moderate." Excluding these two observations, the correlation between the skin cancer

9

incidence rate and the *UV Intensity* is only .021. The cloud of points in Figure 1 underscores the finding that there is no (let alone positive) association between the two variables.

Suppose we divide our sample of fifty-seven cities into two groups, those above the 37th parallel (n_1 = 37) and those below the 37th parallel (n_2 = 20). From east to west, the 37th parallel extends from about Norfolk, Virginia, to San Francisco, California, and runs along the northern border of North Carolina to the northern border of Arizona, including most of California. One would expect a higher skin cancer incidence rate below the 37th parallel than above it. In fact, there is *no* statistically discernible difference between the two average rates (average above = 20.82, average below = 18.94, *p* = .212 for a two-tailed test). Surprisingly, the average rate is marginally higher above the 37th parallel which is associated with *less* sun exposure.

The scatterplot in Figure 2, like Figure 1, shows the relationship between skin cancer incidence rates and *UV Intensity*. But, diamonds are used to denote cities that lie below the 37th parallel, while dots denote cities that lie above the 37th parallel. Note that the points in the lower right-hand part of the scatterplot (that is, cities with 50 percent or more days of the year having a UV Index forecast of "High," "Very High" or "Extreme") have relatively *low* skin cancer incidence rates (that is, *all* of these points are diamonds). Note that the points in the upper left- hand part of the scatterplot (that is, cities with less than 30 percent of all days of the year having a UV Index forecast of "High" or higher) have among the highest skin cancer incidence rates (that is, *all* of these points are dots). The scatterplot in Figure 2 suggests that there might be strong regional differences.

The fifty-five observations on cities in the continental U.S. were divided into four census regions: West (with state abbreviations of CA, OR, WA, ID, MT, WY, NV, UT, CO, AZ, NM), Midwest (ND, SD, NE, KS, MO, IA, MN, WI, IL, IN, OH, MI), Northeast (PA, NJ, NY, CT, RI, MA, VT, NH, ME), and South (TX, OK, LA, AR, MS, AL, TN, KY, FL, GA, SC, NC, VA, WV, DE, MD, and the District of Columbia). Table 2 summarizes the results of two sets of 2-sample *t*-tests, one involving average *UV Intensity* between census regions and the other involving average skin cancer incidence rates between census regions. The results appearing in the top half of Table 2 are not at all surprising: the South and West are the two sunniest regions. The results appearing in the bottom half of Table 2 are, however, *very* surprising. Skin cancer incidence rates are, on average, highest in the Northeast and significantly higher there than they are in either the South (*p* = .042) or in the Midwest (*p* = .005). The map of the continental U.S. in Figure 3 underscores the strong regional differences observed in Table 2.[4,5]

If the sun's radiation is most intense in the South, then why are skin cancer incidence rates highest in the Northeast? Dr. Michael F. Holick [**1**, pp. 33] writes that "genetics plays a much more important role in the development of melanoma than does regular, moderate sun exposure." Moreover, he points out that melanomas typically occur on parts of the body that receive no sun exposure. And, finally, residents in the South are more likely to build a resistance to sunburn in the form of tanned skin. Tanned skin, in turn, can protect against sunburn, thought to be the main cause of melanoma.

Concluding Remarks

Some people are convinced that no amount of sun exposure is safe. The less, the better. The results presented here, however, show otherwise. In the region of the country where sun exposure is most intense, the most dangerous form of skin cancer, melanoma, is less of a problem than it is in regions of the country with far less sun exposure.

An examination of three-year average incidence rates of melanoma and other non-epithelial skin cancers (excluding basal and squamous) for 55 metropolitan statistical areas in the continental U.S. (2006 through 2008, and the latest available data at the MSA level as of 2012) shows that people who live in sunny climates have a *lower* risk of melanoma than do people who live in climates with only limited amounts of sunlight.

In the words of The Beatles, "here comes the sun, and I say it's all right."

Table 1. Three-Year Average Age-Adjusted Skin Cancer Rates and

Daily UV Index Forecast in Selected Cities,

2006-2008

City	Latitude[1]	Skin Cancer Rate[2]	UV Index Forecast[3]				
			Low	Moderate	High	Very High	Extreme
Albuquerque, NM	35.084	23.353	145	309	171	325	111
Anchorage, AK	61.218	11.537	785	276	0	0	0
Atlanta, GA	33.749	26.712	241	303	167	349	1
Atlantic City, NJ	39.364	20.050	441	250	200	170	0
Baltimore, MD	39.290	25.545	430	246	211	174	0
Billings, MT	45.783	20.191	468	254	124	215	0
Bismarck, ND	46.808	19.869	516	263	159	123	0
Boise, ID	43.614	26.094	395	269	151	246	0
Boston, MA	42.358	25.809	479	248	194	140	0
Buffalo, NY	42.886	21.329	505	244	236	76	0
Burlington, VT	44.476	28.828	577	243	177	64	0
Charleston, SC	32.776	24.411	186	339	153	376	7
Charleston, WV	38.350	19.987	455	241	162	203	0
Cheyenne, WY	41.140	20.928	346	283	144	273	15
Chicago, IL	41.850	14.658	485	265	180	131	0

Cleveland, OH	41.499	18.448	483	254	198	126	0
Concord, NH	43.208	28.216	523	238	193	107	0
Dallas, TX	32.783	14.136	230	324	146	351	10
Denver, CO	39.739	20.728	304	274	141	307	35
Des Moines, IA	41.601	18.950	474	268	157	162	0
Detroit, MI	42.331	18.829	513	237	210	101	0
Dover, DE	39.158	20.050	439	253	200	169	0
Hartford, CT	41.764	26.804	493	243	190	135	0
Honolulu, HI	21.307	17.339	15	102	244	332	368
Houston, TX	29.763	15.678	130	320	157	407	47
Indianapolis, IN	39.768	20.433	454	256	174	177	0
Jackson, MS	32.299	18.003	187	319	159	377	19
Jacksonville, FL	30.332	19.471	100	328	164	414	55
Las Vegas, NV	36.175	14.613	216	288	170	372	15
Little Rock, AR	34.746	13.430	309	261	162	329	0
Los Angeles, CA	34.052	19.878	147	304	149	343	118
Louisville, KY	38.254	23.644	420	247	177	217	0
Memphis, TN	35.149	14.506	330	268	161	302	0
Miami, FL	25.774	16.420	43	230	168	348	272
Milwaukee, WI	43.039	18.421	494	253	218	96	0
Minneapolis, MN	44.980	0.388	529	242	200	90	0
Mobile, AL	30.694	20.375	133	309	156	404	59

Table 1. Three-Year Average Age-Adjusted Skin Cancer Rates and

Daily UV Index Forecast in Selected Cities,

2006-2008

(Continued)

City	Latitude[1]	Skin Cancer Rate[2]	UV Index Forecast[3]				
			Low	Moderate	High	Very High	Extreme
New Orleans, LA	29.954	13.508	113	311	145	381	111
New York, NY	40.714	18.367	461	260	190	150	0
Norfolk, VA	36.847	22.240	341	279	169	272	0
Oklahoma City, OK	35.468	22.416	300	291	164	303	3
Omaha, NE	41.259	19.886	446	260	159	191	0
Philadelphia, PA	39.952	20.050	455	252	202	152	0
Phoenix, AZ	33.448	17.275	140	307	165	391	58
Pittsburgh, PA	40.441	15.546	500	217	199	145	0
Portland, ME	43.661	30.347	519	253	196	93	0
Portland, OR	45.524	27.967	578	191	157	135	0
Providence, RI	41.824	22.210	500	230	188	143	0
Raleigh, NC	35.772	24.566	327	269	173	292	0
Salt Lake City, UT	40.761	31.285	330	262	142	285	42

City	Latitude						
San Francisco, CA	37.775	22.715	292	261	150	357	1
Seattle, WA	47.606	28.046	572	242	173	74	0
Sioux Falls, SD	43.550	15.661	493	244	163	161	0
St. Louis, MO	38.627	20.398	424	253	149	235	0
Tampa Bay, FL	27.947	20.442	71	290	165	371	164
Washington, DC	38.895	17.038	422	259	190	190	0
Wichita, KS	37.692	11.227	385	279	133	263	1

[1] Each city's latitude is calculated at http://www.travelmath.com/cities/ .

[2] Three-year (2006-2008) skin cancer incidence rates per 100,000 people excluding basal and squamous (but including melanoma and other non-epithelial skin cancers) for all ages, both genders, all ethnicities and all races at http://wonder.cdc.gov/cancer-v2008.HTML . Rates are reported for the metropolitan statistical area (MSA) to which each city belongs.

[3] The daily UV Index forecasts for the three-year period (2006-2008) are from http://www.cpc.ncep.noaa.gov/products/stratosphere/uv_index/index.html . Annual time series for UV Index forecasts are reported for each of the fifty-seven cities. The observations on the five categories do not add up to 3×365 due to missing data.

Table 2. Summary of Two-Sample *t*-Tests on

Average UV Intensity and Skin Cancer Rates (2006-2008)

Between Census Regions

UV Intensity[1]

		Average		*p*-value on difference
Group 1	Group 2	Group 1	Group 2	
West[2] (n = 12)	Midwest (n = 12)	43.630	31.092	**.004**
West[2] (n = 12)	Northeast (n = 11)	43.630	30.332	**.003**
West[2] (n = 12)	South (n = 20)	43.630	48.662	.247
Midwest (n = 12)	Northeast (n = 11)	31.092	30.332	.589
Midwest (n = 12)	South (n = 20)	31.092	48.662	**<.001**
Northeast (n = 11)	South (n = 20)	30.332	48.662	**<.001**

Skin Cancer Rate[3]

| | | | Average | | p-value on difference |
Group 1	Group 2	Group 1	Group 2	
West[2] (n = 12)	Midwest[4] (n = 11)	22.756	17.889	**.008**
West[2] (n = 12)	Northeast (n = 11)	22.756	23.414	.748
West[2] (n = 12)	South (n = 20)	22.756	19.629	.078
Midwest[4] (n = 11)	Northeast (n = 11)	17.889	23.414	**.005**
Midwest[4] (n = 11)	South (n = 20)	17.889	19.629	.184
Northeast (n = 11)	South (n = 20)	23.414	19.629	**.042**

[1]UV intensity is represented by the percentage of days the UV Index forecast was "High,"

"Very High" or "Extreme" over the three-year period 2006 through 2008.

[2]Excluding the observations on Honolulu, HI and Anchorage, AK.

[3]Skin cancer incidence rate (excluding basal and squamous) for selected MSAs in each census region over the three-year period 2006 through 2008.

[4]Excluding the observation on the MSA for Minneapolis, MN. Including this observation would accentuate the difference, that is, reduce the p-value of the test, if the Midwest region's average was already lower than the average of the comparison group.

Figure 1. Scatterplot of Skin Cancer Rates and *UV Intensity*

Averaged Over the Three-Year Period 2006-2008,

57 Metropolitan Statistical Areas

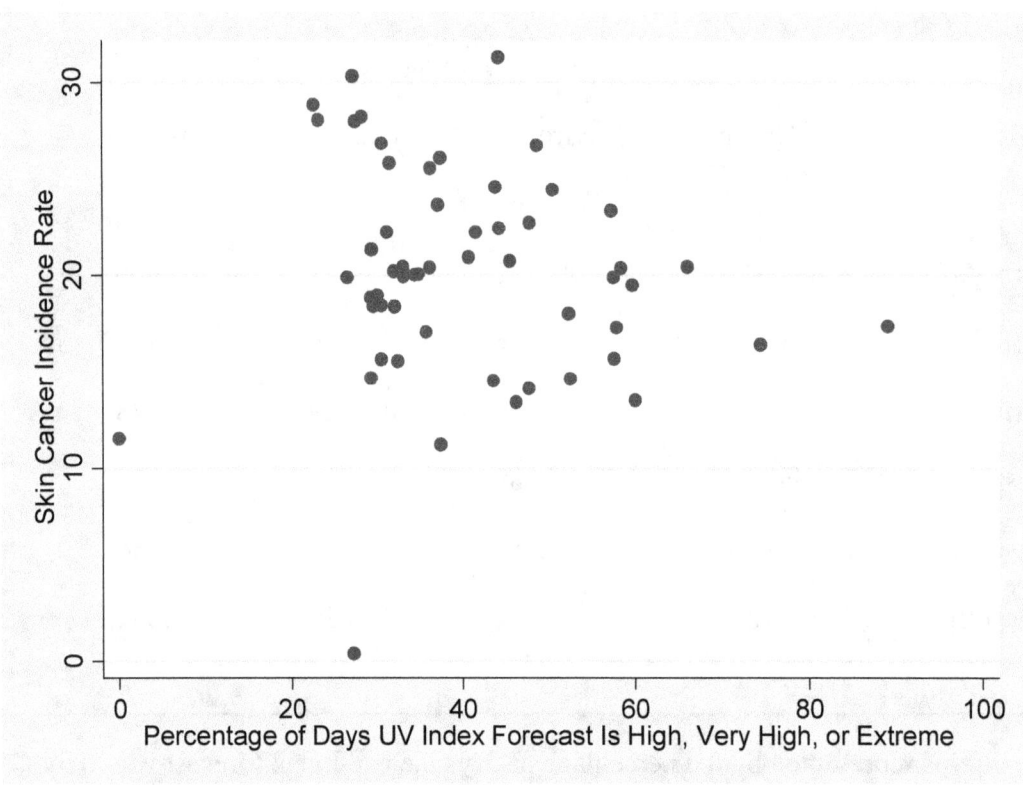

Figure 2. Scatterplot of Skin Cancer Rates and *UV Intensity*

Averaged Over the Three-Year Period 2006-2008,

57 Metropolitan Statistical Areas,

Above or Below the 37th Parallel

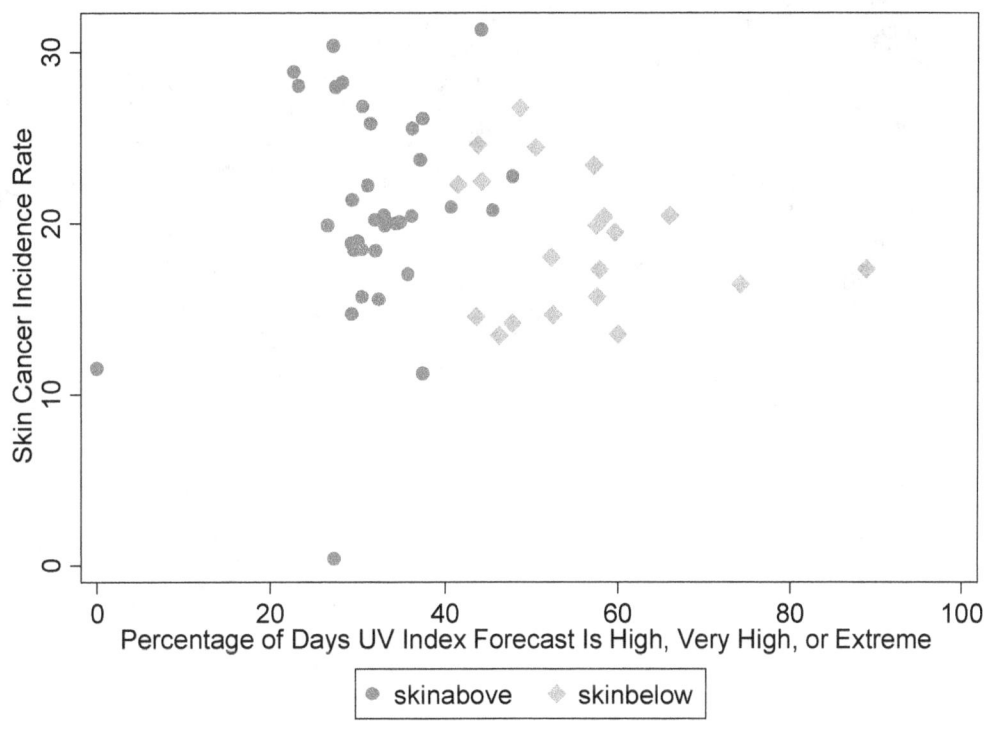

Figure 3. Skin Cancer Incidence Rates (2006-2008) Above and Below the 37th Parallel,

by State and Census Region

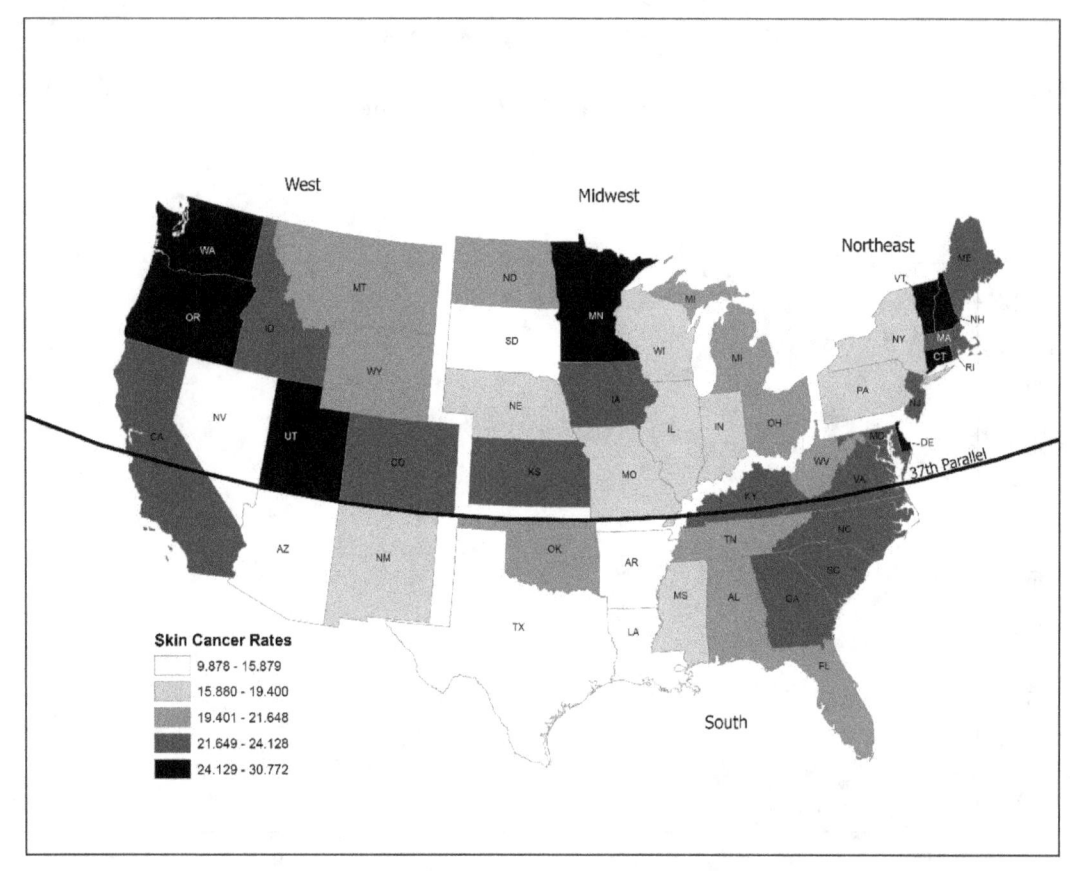

Reference

1. M. F. Holick and M. Jenkins, *The UV Advantage*, iBooks, New York, NY, 2003

Footnotes

1. Certain surfaces (like snow, sand, or water) reflect UV radiation and increase its intensity.

2. As of 2012, the latest rates at the MSA level are for the year 2008.

3. Most of the fifty-seven cities could be matched with one of the CDC's cancer website's MSAs. The exceptions were: Mobile, AL (for which the Birmingham-Hoover, AL MSA was used); Billings, MT (for which the state of Montana was used); Anchorage, AK (the state of Alaska); Bismarck, ND (North Dakota); Wichita, KS (Kansas City, MO-KS); Sioux Falls, SD (South Dakota); Concord, NH (New Hampshire); Atlantic City, NJ and Dover, DE (Philadelphia-Camden-Wilmington, PA-NJ-DE-MD); Burlington, VT (Vermont); Norfolk, VA (Richmond, VA); Charleston, WV (West Virginia); and Cheyenne, WY (Wyoming).

4. The author gratefully acknowledges the assistance of William Hegman, GIS Specialist and Teaching Fellow at Middlebury College.

5. Three-year (2006-08) statewide skin cancer incidence rates are used in Figure 3. Perhaps most noteworthy is the very low skin cancer rate for the Minneapolis-St. Paul MSA (0.388) reported in Table 1, but the very high overall rate reported for the state of Minnesota (25.614) in Figure 3. What are residents of Minneapolis-St. Paul doing differently from the rest of the state to protect themselves against the most dangerous form of skin cancer?

The Apollonius Circles of Rank k

Professor Ion Patrascu
Fratii Buzesti National College
Craiova, Romania

Professor Florentin Smarandache
New Mexico University, Gallup, NM USA

Abstract

The purpose of this article is to introduce the notion of the Apollonius circle of rank k and generalize some results on Apollonius circles.

Definition 1: The line AA_k where $A_k \in (BC)$, such that

$$\frac{BA}{A_k C} = \left(\frac{AB}{AC}\right)^k \quad (k \in \mathbb{R})$$

is called an **internal cevian of rank k**.

If A_k' is the harmonic conjugate of the point A_k in relation to B and C, we call the line AA_k' an **outside cevian of rank *k***.

Definition 2. We call the circle which has diameter the segment line $A_k A_k'$ with respect to the side BC of triangle ABC the **Apollonius circle of rank *k***.

Theorem 1: The Apollonius circle of rank k is the locus of points M from triangle ABC's plan, satisfying the relation:

$$\frac{MB}{MC} = \left(\frac{AB}{AC}\right)^k.$$

Proof: Let O_{A_k} be the center of the Apollonius circle of rank k relative to the side BC of triangle ABC (see figure 1). Let U, V be the points of intersection of this circle with the circle circumscribed to the triangle ABC. We denote by D the middle of arc BC, and we extend DA_k to intersect the circle circumscribed in U'. In triangle $BU'C$, $U'D$ is a bisector; it follows that

$$\frac{BA_k}{A_k C} = \frac{U'B}{U'C} = \left(\frac{AB}{AC}\right)^k,$$

so U' belongs to the locus. The perpendicular in U' on $U'A_k$ intersects BC on A_k'', which is the foot of triangle $BUC's$ outer bisector, so it is the harmonic conjugate of A_k in relation to B and C, thus $A_k'' = A_k'$. Therefore, U' is on the Apollonius circle of rank k relative to the side BC, hence $U' = U$.

Let M be a point that satisfies the relation from the statement of the theorem; thus

$$\frac{MB}{MC} = \frac{BA_k}{A_k C}.$$

It follows – by using the reciprocal of bisector's theorem – that MA_k is the internal bisector of angle BMC. Now let us proceed as before, taking the external bisector; it follows that belongs to the Apollonius circle of center O_{A_k}. We consider now a point M on this circle, and we construct C' such that

$$\sphericalangle BNA_k \equiv \sphericalangle A_k NC'$$

23

(thus NA_k is the internal bisector of the angle BNC'. Because $A'_k N \perp NA_k$, it follows that A_k and A'_k are harmonically conjugated with respect to B and C'. On the other hand, the same points are harmonically conjugated with respect to B and C; from here, it follows that $C' = C$, and we have

$$\frac{NB}{NC} = \frac{BA_k}{A_k C} = \left(\frac{AB}{AC}\right)^k.$$

Figure 1

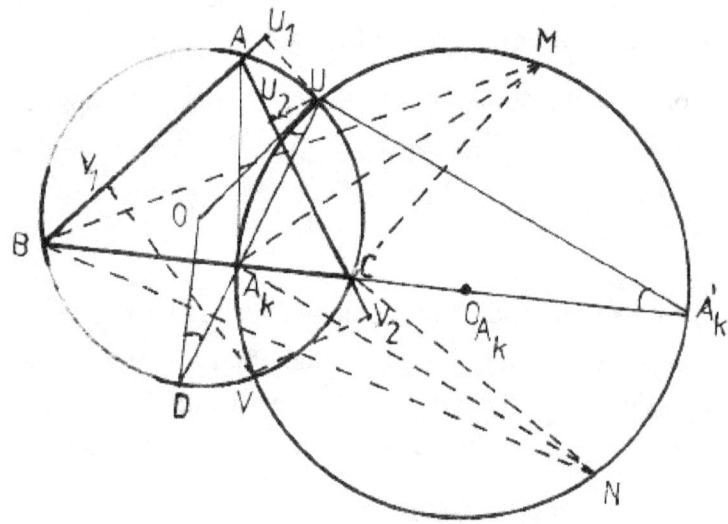

Definition 3: The geometric figure obtained from a convex quadrilateral by extending the opposite sides until they intersect is called a **complete quadrilateral**. A complete quadrilateral has 6 vertices, 4 sides and 3 diagonals.

Theorem 2: In a complete quadrilateral, the three means of the three diagonals are collinear (Gauss - 1810).

Proof: Let $ABCDEF$ a given complete quadrilateral (see figure 2). We denote by H_1, H_2, H_3, H_4 respectively the orthocenters of triangles ABF, ADE, CBE, CDF, and let A_1, B_1, F_1 be the feet of the heights of triangle ABF.

As previously shown, the following relations occur

$$H_1 A * H_1 A_1 - H_1 B * H_1 B_1 = H_1 F * H_1 F_1$$

they express that the point H_1 has equal powers to the circles of diameters AC, BD, EF, because those circles contain respectively the points A_1, B_1, F_1, and H_1 is an internal point. It is shown analogously that the points H_2, H_3, H_4 have equal powers to the same circles, so those points are situated on the radical axis (common to the circles), therefore the circles are part of a fascicle, as

24

such their centers – which are the means of the complete quadrilateral's diagonals – are collinear. The line formed by the means of a complete quadrilateral's diagonals is called the **Gauss** or **Gauss-Newton line**.

Figure 2

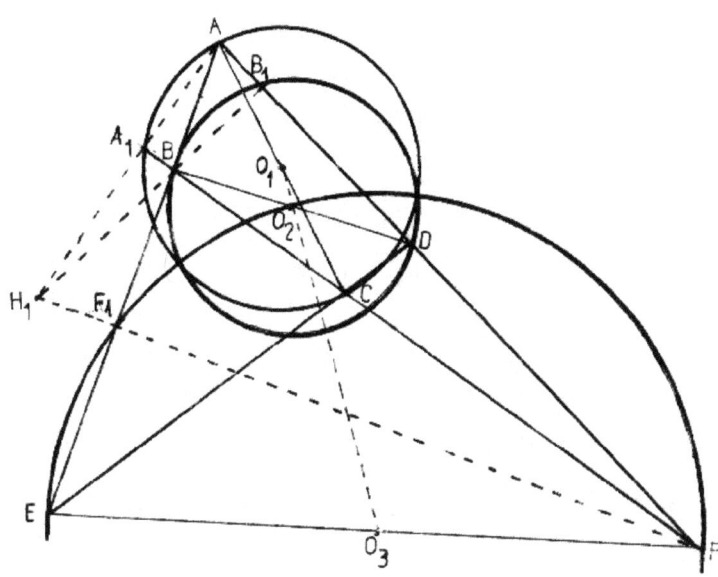

Theorem 3: The Apollonius circles of rank k of a triangle are part of a fascicle.

Proof: Let AA_k, BB_k, CC_k be concurrent cevians of rank k and AA'_k, BB'_k, CC'_k be the external cevians of rank k (see figure 3). The figure $B'_k C_k B_k C'_k A_k A'_k$ is a complete quadrilateral so theorem 2 can be applied.

Theorem 4: The Apollonius circles of rank k of a triangle are the orthogonals of the circle circumscribed to the triangle.

Proof: We unite O to D and U (see figure 1), $OD \perp BC$ and $m(A_k U A'_k) = 90^0$, it follows that $\widehat{UA'_k A_k} = \widehat{ODA_k} = \widehat{OUA_k}$. The congruence $\widehat{UA'_k A_k} \equiv \widehat{OUA_k}$ shows that OU is tangent to the Apollonius circle of center O_{A_k}. Analogously, this can be demonstrated for the other Apollonius circles.

Remark 1: The previous theorem indicates that the radical axis of Apollonius circles of rank k is the perpendicular taken from O to the line $O_{A_k} O_{B_k}$.

Figure 3

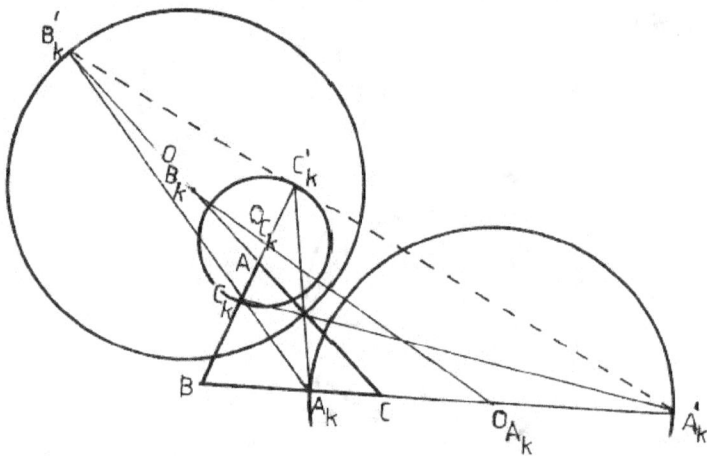

Theorem 5: The centers of Apollonius circles of rank k of a triangle are situated on the trilinear polar associated to the intersection point of the cevians of rank $2k$.

Proof: From the previous theorem, it follows that $OU \perp UO_{A_k}$, so UO_{A_k} is an external cevian of rank 2 for the triangle BCU, thus an external symmedian. Henceforth,

$$\frac{O_{A_k}B}{O_{A_k}C} = \left(\frac{BU}{CU}\right)^2 = \left(\frac{AB}{AC}\right)^{2k}$$

(the last equality occurs because U belongs to the Apollonius circle of rank k associated to the vertex A).

Theorem 6: The Apollonius circles of rank k of a triangle intersect the circle circumscribed to the triangle in two points that belong to the internal and external cevians of rank $k + 1$.

Proof: Let U and V be the points of intersection of the Apollonius circle of center O_{A_k} with the circle circumscribed to the ABC (see figure 1). We take from U and V the perpendiculars UU_1, UU_2 and VV_1, VV_2 on AB and AC respectively. The quadrilaterals $ABVC$, $ABCU$ are inscribed, the similarity of triangles BVV_1, CVV_2 and BUU_1, CUU_2 follows, from which we get the relations:

$$\frac{BV}{CV} = \frac{VV_1}{VV_2}, \qquad \frac{UB}{UC} = \frac{UU_1}{UU_2}.$$

Furthermore,

$$\frac{BV}{CV} = \left(\frac{AB}{AC}\right)^k, \frac{UB}{UC} = \left(\frac{AB}{AC}\right)^k, \frac{VV_1}{VV_2} = \left(\frac{AB}{AC}\right)^k \text{ and } \frac{UU_1}{UU_2} = \left(\frac{AB}{AC}\right)^k,$$

26

relations that show that V and U belong respectively to the internal cevian and the external cevian of rank k + 1.

Definition 4: If the Apollonius circles of rank k associated with a triangle have two common points, then we call these points isodynamic points of rank k (and we denote them W_k, W_k').

Property 1: If W_k, W_k' are isodynamic centers of rank k, then:

$$W_k A * BC^k = W_k B * AC^k = W_k C * AB^k; \; W_k'A * BC^k = W_k'B * AC^k = W_k'C * AB^k.$$

The proof of this property follows immediately from theorem 1.

Remark 2: The Apollonius circles of rank 1 are the investigated Apollonius circles (the bisectors are cevians of rank 1). If k = 2, the internal cevians of rank 2 are the symmedians, and the external cevians of rank 2 are the external symmedians, i.e. the tangents in the triangle's vertices to the circumscribed circle. In this case, for the Apollonius circles of rank 2, theorem 3 is modified to:

Theorem 7: The Apollonius circles of rank 2 intersect the circumscribed circle to the triangle in two points belonging respectively to the antibisector's isogonal and to the cevian outside of it.

The proof of this theorem follows from the proof of theorem 6. We mention that the antibisector is isotomic to the bisector, and a cevian of rank 3 is isogonic to the antibisector.

References

1. N. N. Mihăileanu: **Lecţii complementare de geometrie**
[*Complementary Lessons of Geometry*], Editura Didactică şi Pedagogică, Bucureşti, 1976.

2. C. Mihalescu: **Geometria elementelor remarcabile**
[*The Geometry of Outstanding Elements*], Editura Tehnică, Bucureşti, 1957.

3. V. Gh. Vodă: **Triunghiul – ringul cu trei colţuri**
[*The Triangle – The Ring with Three Corners*], Editura Albatros, Bucureşti, 1979.

4. F. Smarandache, I. Pătraşcu: **Geometry of Homological Triangle**, The Education Publisher Inc., Columbus, Ohio, SUA, 2012.

A TALE OF TWO … UMM THREE ZITIS

Casey Y. Park
Theodora L. Yoch
Eleanor G. Kaufman
Paul M. Sommers

Department of Economics
Middlebury College
Middlebury, Vermont 05753
psommers@middlebury.edu

Abstract

Choosing a restaurant is only one click away, thanks to numerous restaurant rating websites. Zagat uses a 30-point scale to rate food quality, décor, and service to help diners decide where to eat. Using a series of two-sample t-tests, the authors examine Italian food dining experience in three U.S. cities with large numbers of Italian Americans – Boston, Chicago, and Philadelphia. The median cost is used to divide each of the three samples into less and more expensive Italian restaurants. Overall, the results suggest that Chicago has the highest average ratings in all three categories.

"It was the best of times,
it was the worst of times."
— Charles Dickens,
A Tale of Two Cities [1]

Everyone loves to eat out. But, the dining experience can be the best of times —
or the worst of times. To help take the guesswork out of where to eat, we frequently take a look
at restaurant ratings by diners. The Zagat Survey was established by Tim and Nina Zagat in
1979 to collect the ratings of restaurants by diners in New York City. The Zagat Survey now
includes fifteen major U.S. cities, three of which are examined in this brief research note:
Boston, Chicago, and Philadelphia. For all three cities, we evaluate their Italian restaurants.
In May 2015, data on food quality, décor, and service (with ratings on a 30-point scale, the
higher the rating, the stronger the recommendation) were collected from 74 Italian restaurants in
Boston, 90 in Chicago, and 84 in Philadelphia.[1,2] For each of the three selected cities, the sample
was then divided into two groups, less expensive restaurants whose cost was less than or equal to
the median cost in that particular city and more expensive restaurants whose cost was above the
median. For restaurants, the "cost" includes the price of dinner with one drink and tip.[3] A series
of two-sample *t*-tests compared the difference between two means, namely, the mean rating (of
food quality, décor, or service) in one city to the corresponding mean rating in another city for
each of three groups: (i) all Italian restaurants; (ii) less expensive Italian restaurants; and
(iii) more expensive Italian restaurants.
The results of the various two-sample *t*-tests are summarized in Tables 1, 2, and 3. The
dimensional ratings are as follows: 26 – 30, extraordinary to perfection; 21 – 25, very good to
excellent; and 16 – 20, good to very good.[4] Table 1 summarizes the pairwise comparisons
between any two of the three selected cities on the basis of food quality.[5] For all restaurants,
food quality, on average, was higher in Chicago than in Philadelphia ($p = .003$), but there was no
difference in average food quality between Boston and Philadelphia ($p = .175$) or between
Boston and Chicago ($p = .238$).
The median cost (dividing "less expensive" from "more expensive" restaurants) in Boston,
Chicago, and Philadelphia was $40.00, $38.50, and $38.50, respectively. All three median cost
figures fall into Zagat's moderate ($26 - $40) range. The middle third of Table 1 reveals that
less expensive Italian restaurants in Chicago enjoyed a significantly higher average food quality
than did their counterparts in Philadelphia ($p = .003$). The bottom third of Table 1 reveals no
differences in average food quality among more expensive restaurants between any two of the
three selected cities.
The results of comparisons for décor are summarized in Table 2. For all Italian restaurants,
Chicago enjoys the highest average rating, even when the sample is divided into less expensive
restaurants. Again, among more expensive restaurants, Chicago's average décor rating is
highest, higher than Philadelphia's average ($p = .024$) but not discernibly different from Boston's
($p = .231$).

Finally, Table 3 summarizes the results for service at Italian restaurants. Among the less expensive restaurants, Chicago emerges as the best of the three selected cities, although no differences appear among the more expensive restaurants.

Concluding Remarks

Zagat ratings at Italian restaurants in three U.S. cities — Boston, Chicago, and Philadelphia — with large populations of Italian Americans are examined for differences in average food quality, décor, and service. Chicago's food quality is best, significantly better than the average food quality in Philadelphia (which probably built a stronger reputation for its cheesesteak than for its *formaggio*). Among less expensive Italian restaurants, Chicago scores significantly higher in décor and service than its counterparts in either Boston or Philadelphia. In short, we conclude that among these three cities, Chicago provides the best overall dining experience at Italian restaurants.

Sidney Carton, a central character in Dickens' *A Tale of Two Cities*, might never have worried about food quality of Italian restaurants in Paris or London (let alone décor or service, unless Lucie Manette was his waitress). But, if he had found himself in the Windy City during the last few days of his life and was looking for particularly good Italian cuisine, his unspoken last thoughts might have been:

> It is a far, far better thing that I do,
> than I have ever done. It is a far,
> far better rest(aurant) that I go to
> than I have ever known.

Table 1. Ratings Comparisons of Italian Restaurants by Food Quality, Between Selected Cities

All Restaurants

City 1	City 2	Averages City 1	Averages City 2	p-value on difference*
Boston (n = 74)	Philadelphia (n = 84)	24.135	23.738	.175
Boston (n = 74)	Chicago (n = 90)	24.135	24.444	.238
Philadelphia (n = 84)	Chicago (n = 90)	23.738	24.444	**.003**

Less Expensive Restaurants

City 1	City 2	Averages City 1	Averages City 2	p-value on difference
Boston (n = 38)	Philadelphia (n = 42)	23.842	23.286	.141
Boston (n = 38)	Chicago (n = 45)	23.842	24.267	.166
Philadelphia (n = 42)	Chicago (n = 45)	23.286	24.267	**.003**

More Expensive Restaurants

City 1	City 2	Averages City 1	Averages City 2	p-value on difference
Boston (n = 36)	Philadelphia (n = 42)	24.444	24.190	.562
Boston (n = 36)	Chicago (n = 45)	24.444	24.622	.677
Philadelphia (n = 42)	Chicago (n = 45)	24.190	24.622	.192

*p-value on two-sample two-tailed t-test.

**Table 2. Ratings Comparisons of Italian Restaurants by Décor,
Between Selected Cities**

All Restaurants

| City 1 | City 2 | Averages | | p-value |
		City 1	City 2	on difference*
Boston	Philadelphia	20.162	19.821	.460
(n = 74)	(n = 84)			
Boston	Chicago	20.162	21.056	**.017**
(n = 74)	(n = 90)			
Philadelphia	Chicago	19.821	21.056	**.002**
(n = 84)	(n = 90)			

Less Expensive Restaurants

| City 1 | City 2 | Averages | | p-value |
		City 1	City 2	on difference
Boston	Philadelphia	18.921	18.738	.749
(n = 38)	(n = 42)			
Boston	Chicago	18.921	20.044	**.011**
(n = 38)	(n = 45)			
Philadelphia	Chicago	18.738	20.044	**.017**
(n = 42)	(n = 45)			

More Expensive Restaurants

| City 1 | City 2 | Averages | | p-value |
		City 1	City 2	on difference
Boston	Philadelphia	21.472	20.905	.366
(n = 36)	(n = 42)			
Boston	Chicago	21.472	22.067	.231
(n = 36)	(n = 45)			
Philadelphia	Chicago	20.905	22.067	**.024**
(n = 42)	(n = 45)			

*p-value on two-sample two-tailed t-test.

Table 3. Ratings Comparisons of Italian Restaurants by Service, Between Selected Cities

All Restaurants

City 1	City 2	Averages		*p*-value
		City 1	City 2	on difference*
Boston	Philadelphia	22.311	22.321	.971
(n = 74)	(n = 84)			
Boston	Chicago	22.311	22.667	.186
(n = 74)	(n = 90)			
Philadelphia	Chicago	22.321	22.667	.179
(n = 84)	(n = 90)			

Less Expensive Restaurants

City 1	City 2	Averages		*p*-value
		City 1	City 2	on difference
Boston	Philadelphia	21.763	21.595	.637
(n = 38)	(n = 42)			
Boston	Chicago	21.763	22.444	**.033**
(n = 38)	(n = 45)			
Philadelphia	Chicago	21.595	22.444	**.016**
(n = 42)	(n = 45)			

More Expensive Restaurants

City 1	City 2	Averages		*p*-value
		City 1	City 2	on difference
Boston	Philadelphia	22.889	23.048	.707
(n = 36)	(n = 42)			
Boston	Chicago	22.889	22.889	1.000
(n = 36)	(n = 45)			
Philadelphia	Chicago	23.048	22.889	.646
(n = 42)	(n = 45)			

*p-value on two-sample two-tailed t-test.

References

1. Dickens, Charles. *A Tale of Two Cities*. New York: Modern Library, 1935.

2. "What do Zagat Ratings Mean?" at
    https://support.google.com/Zagat/answer/1705271?hl/=en .

Footnotes

1. According to the 1990 United States Census, the four cities with the most Italian
    Americans are New York City, NY (1,882,396); Philadelphia, PA (497,721);
    Chicago, IL (492,158); and Boston, MA (485,761). Population figures include
    the greater metropolitan area. See https://www.niaf.org/culture/statistics/5187-2/ .

2. See https://www.zagat.com/boston, https://www.zagat.com/chicago, and
    https://www.zagat.com/philadelphia, all of which were accessed between
    May 20th and May 25th 2015. For each city, click on the box for "Italian"
    cuisines for a list of all Italian restaurants.

3. See [2] at https://support.google.com/zagat/answer/1705271?hl=en.

4. *Ibid.*

5. New York City, with the highest number of Italian Americans and the largest
    number of Italian restaurants in the Zagat Survey, 371, had an average food
    quality rating of 22.294, an average décor rating of 18.388, and an average
    service rating of 20.725. In comparisons of New York City with Boston, Chicago,
    and Philadelphia, the p-value on the two-tailed alternative was in all cases less than or equal to
    .0001. That is, the average rating of food quality, décor, or service was discernibly lower in
    New York City than the corresponding average in any of the three other cities.

ON THE NON- RANDOMNESS OF THE BEAST NUMBER

N.E.Myridis

Ass.Professor

Aristotle University of Thessaloniki

Thessaloniki, Greece 54124

e-mail: nmyridis@theo.auth.gr

Abstract

This paper relates abstract studies with mathematical research. This work falls into the field of the arithmetic of revelation; moreover it relates to –maybe- the most curious number of history, i.e. 666. The present paper reveals the non-randomness of this number regarding its unique reference in the Bible in the book of Revelation.

Introduction

The mathematical field related to the arithmetic of Revelation has already been introduced [1]. The numbers appearing in the text of Revelation in the New Testament are miscellaneous. Without any doubt the most famous and curious number is that of 666 (in the original (Greek) text this number is referred to as $\chi\xi\varsigma'$)[1]. The notoriety of this number goes beyond the borders of religious areas as well as of historical affairs. It has been a part and logo of our daily life for many decades (see for instance [2], [3] or even the term *hexakosioihexikontaexphobia*[2]), almost in a worldwide perspective. We should also state that we could claim that $\chi\xi\varsigma'$ is the most curious number in history. In the following sections we further examine this number using the point of view of coding it in different arithmetic systems. As a consequence we will conclude a really prime and paradoxical result: the hidden and absolute relation of the number 666 to its coordinates in the Bible.

Revelation numbers

The numbers appearing in the Book of Revelation (New Testament) may be referred to as *'revelation numbers'*. The most well known of these numbers is 666[3], which is usually called the *number of the beast*. It is well known that this mathematical subject, known as *the Arithmetic of Revelation* [1], has generated particular debate and study. Examples include several diligent works (mathematical (e.g. [4], [5]) and theological (e.g. [6], [7], [8])). However, beyond the full scientific investigation and approach of the topic, we must note that the use of the specific number and the expansion of its impacts, often takes on a mythical nature. Which, accordingly gives rise to divergent and vague claims and findings. The result of all these procedures consists of the certainty that the philology about the specific number appears as a mysterious and unknown area.

Nevertheless, the involvement of mathematical and computational studies and their respective research with revelation numbers is an area with a specific degree of surprise. In fact, the extensive synopsis of the basic findings and mathematical characteristics of these numbers may be found in mathematical references [9]. It is worth mentioning that in mathematical circles, terms are used which refer to the specific number (666) and to its affinities. Examples include the terms used in Table 1. We note that the relative analysis falls into the field of number theory. At this point, we present the characteristic identity of each apocalyptic number, which gives them their particular name.

1. *Beast number.* It is the well known number of the Apocalypse, *666*, or more accurately the Greek equivalent $\chi\xi\varsigma'$. The properties of this number will be presented in the next section.

---

[1] The citation of this number in the New Testament only happens once in (Revelation 13:18).
[2] That is 'the fear of 666'.
[3] "Here is wisdom. Let him who has understanding calculate the number of the beast, for it is the number of a man: his number *is* 666." (Revelation 13:18)

2. *Repdigit.* When a number, in any numerical basis, consists of similar digits, which are repeated. 666 is a "repdigit" (repeated digit).
3. *Apocalypse number.* A number of the form 2^i that contains the beast number within its digits.
4. *Apocalyptic number.* A number that has 666 decimal digits. Also, this category includes numbers that have even number of 1s in their binary representation.
5. *Bimonster.* The number which is the wreathed product of the monster group by Z_2. If the respective parameters are positioned with one next to the other, this number can be denoted by 666 and corresponds somehow to the diagram inside table 1.
6. *Evil number.* The number in which the first n decimal digits add up to 666.

Also, related to the revelation numbers are the Roman numerals[4], these numerals became correlated with and reduced in the beast number.

In the Roman numbering system, the beast number emerges from all the discrete symbols of this system, arranged one next to the other, only one time, with the exception of the symbol M (=1000) [5], i.e.: *666=DCLXVI.*

The entries in the bibliography also contain numerous interesting attributes of the beast number.

Table 1

Terms and values of mathematical entities, which are related to 666

Terms	Value
Beast number [10]	666
Sign of the devil [4]	666
Repdigit [11]	1, ...11, ...111, ...666...
Apocalypse number	2^i with '666' among the digits
Apocalyptic number	666 decimal digits
Bimonster [12]	
Monster Group [13]	$2^{46} \cdot 3^{20} \cdot 5^9 \cdot 7^6 \cdot 11^2 \cdot 13^3 \cdot 17 \cdot 19 \cdot 23 \cdot 29 \cdot 31 \cdot 41 \cdot 47 \cdot 59 \cdot 71,$
Evil number [14]	*The first n decimal digits adding up to 666*

It is worth noting that the probability, (p_n) that the digits of a number sum to a large number n is given by the formula found in [14]:

---

[4] Roman numerals are a system of numbers used by the Romans. In this system, latin symbols are used to yield a numerical basis (i.e. 5, 10, 50). Consequently, any number can be created using a combination of these symbols.

$$p_1 = 1/9$$

$$
p_n = \begin{cases}
\dfrac{1}{9}\left(1 + \displaystyle\sum_{k=1}^{n-1} p_k\right) & \text{for } n < 10 \\[3em]
\dfrac{1}{9}\left(\displaystyle\sum_{k=1}^{9} p_{n-k}\right) & \text{for } n \geq 10
\end{cases}
$$

and the graphical representation of the corresponding probability density distribution (function) is given in figure 1.

In the case of the number 666, this probability is equal to the number

$$p_{666} \approx \frac{1}{5} - 2.1662 \times 10^{-64}$$

i.e. it is practically equal to 20%!

This observation can constitute the criterion for the evaluation of the variety of conjectures, which from time to time in the past have appeared concerning 666. Such conjectures try to link the indication of St. John's Book of Revelation about the number of the beast, with specific names or numbers[5].

Figure 1

The probability density distribution for the case of the digits' summation of one real number to the number n.

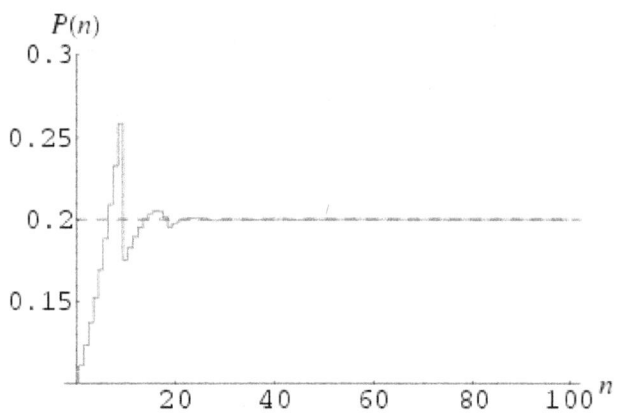

For the beast number (666) and its related expressions there exists a rich bibliography, which concerns the conclusions of the modern mathematical research[6].

[5] Indicatively to this issue we cite the website: http://www.aloha.net/~mikesch/666.htm
[6] An example on this website: http://users.aol.com/s6sj7gt/mike666.htm

Completing the discussion we cite one of our findings regarding the curious number 666, in addition to everything that has been previously cited; i.e., it holds that

$$\frac{6^4}{666} = 1.94594594594595$$

We observe therefore that this ratio is a periodic decimal number with 4 periods equal to 945. The digits of the period add up to 6+6+6! [7]

Thus, this particular symmetrical number (666) has noteworthy and curious properties, which make it especially contemporary and ready-to-use for any group of people or any system.

Observations & Results

Moreover, let us mention a few observations regarding the basis of the arithmetic system that is used. In the previous sections, the arithmetic base used was the number 10. For different arithmetic bases, Table 2 contains results from fundamental calculations.

Observing the values of table 2 it is clear that the number 666 reveals a greater symmetry regarding the other arithmetic systems, except the case of reversing the base and the digit of repetition, i.e. the number 888 reveals greater symmetry ($(888)_6$=4040, $(666)_8$=1232). However we observe that the arithmetic basis 6 creates a greater symmetry.

Table 2

The arithmetic values *666* and *888* in a variety of arithmetic systems

Expression in the decimal number system	Equivalent value in non-decimal system	The arithmetic base
$(666)_{10}$	$(666)_6$=3030	n=6
$(888)_{10}$	$(888)_6$=4040	n=6
$(100)_{10}$	$(100)_6$=244	n=6
$(100)_{10}$	$(100)_8$=144	n=8
$(888)_{10}$	$(888)_8$=1570	n=8
$(666)_{10}$	$(666)_8$=1232	n=8
$(888)_{10}$	$(888)_{11}$=738	n=11
$(666)_{10}$	$(666)_{11}$=556	n=11

Figure 2

---

[7] Moreover, the number 6^4=1296 provides even more interesting symmetrical operations and results, such as

$$6^4*6=7776$$

The latter is a number with a twofold expression which contains three 6-ares and three 7-ares.

An instance of the programming application launched for calculations in different arithmetic systems.

111	55	1	27	1	2
888	111	0	13	7	8
666	83	2	10	3	8
888	148	0	24	4	6
333	111	0	37	0	3
666	222	0	74	0	3
69	11	3	1	5	6
777	111	0	15	6	7
1000	166	4	27	4	6
1000	125	0	15	5	8
1000	1666	4	277	4	6

For the sake of our investigation we launched a relative algorithmic application for calculations in different arithmetic systems. An instance of this application is shown in figure 2.

An Outstanding Observation

At this point, we use the corollaries, notes and applications of the preceding text in order to produce new knowledge regarding the curious number (666). For this purpose we focus on the arithmetic system which uses the base 24. The results of the relevant analysis makes clear that the number of the beast expressed in the base-24 arithmetic system, i.e. $(666)_{24}$ is equal to the number 13[18] (in an alphanumeric arithmetic system)[8]; or equivalently

$$(666)_{24}=13[18]=13[I]=DI$$

or, using the Greek style for the numbers, we could write it in the form:

$$(\chi\xi\varsigma')_{24}=\iota\gamma'[\iota\eta']$$

We use the name $\chi\xi\varsigma'$ *equation* for this equation.

However, as we have already introduced and is known, the citation of the beast number in the Book of Revelation only happens once: in Chapter 13, Verse 18.

The full expansion of the beast number using the base-24 system is:

---

[8] For numbering systems whose base exceeds the ten single-digit numbers (0,…,9), digits from the Latin alphabet are usually used for the rest of the base numbers. In this specific study we use alternative symbols from the Greek numbering system (i.e. 18 is expressed as 'ιη').

$$(666)_{10} = 1 \times 24^2 + 3 \times 24^1 + 18 \times 24^0$$

Obviously, the reason for the use of the base-24 arithmetic system is understandable. The number 666 is introduced into the New Testament with its expression in the Greek numbering system (i.e. 666 ≡ χξϛ'). This system, of course, uses symbols from the Greek alphabet; the latter consists of 24 symbols. Consequently, the reasoning and reference of the number 666 must be found in such kinds of numerical expressions.

Conclusion

The subject of this paper is the investigation of the *inner truth* of the famous number 666; a number with theological origin and hue, as well as with numerous references and implications. Following a kind of literature review regarding this curious number and its affiliates, we present a clear analysis of 666 in different arithmetic systems in an algorithmic way. The results are impressive. There is a unique relation between the point of citation (only once) of this number into the New Testament and the number itself. This relation is present ('encrypted') only in the 24-base arithmetic system.

Thus the aforementioned relation rises even more and becomes a 'strong' relation thanks to the equivalence of the arithmetic base (24) to the number of elements of the arithmetic system which is used to express 666 (i.e. Greek numerals). The overall analysis of the curious number 666 reveals the uniqueness of its use in historical texts. Thus, this analysis indicates the way for the truth that everyone seeks.

References

1. Myridis, N.E. "The Arithmetic of Revelation", *Journal of Recreational Mathematics,* Vol.35(2), 141-142, 2006 (@2009).
2. Heinlein, R.A. *The number of the beast,* Fawcett Publ., 1986.
3. Bolaño, R. *2666,* Barcelona: Editorial Anagrama, 2004.
4. Wang, S. C. "The Sign of the Devil…and the Sine of the Devil." *J. Rec. Math.* 26**,** 201-205, 1994.
5. Wells, D. *The Penguin Dictionary of Curious and Interesting Numbers.* Middlesex, England: Penguin Books, 1986.
6. (a) Andrew of Caesarea, *On the Apocalypse.* In: *The Patrologiae Cursus Completus, Series Graecae,* J.-P.Migne, vol.106, 340 c., Paris, 1857-1866. [prototype]
   (b) Andrew of Caesarea, *Commentary on the Apocalypse (Fathers of the Church).* Cath.Univ.of America Press, 2011. [translation]
7. Beale, G.K. *The Book of Revelation,* Eerdmans Publ., 1998.
8. Gamber, Kl. *Das Geheimnis der dieben Sterne: Zur Symbolik der Apocalypse.* Regensburg, 1987.
9. e.g. http://mathworld.wolfram.com/BeastNumber.html This company is the creator of the well known mathematical software *Mathematica*®.
10. Weisstein, E.W. "Beast Number." From MathWorld—A Wolfram Web Resource. http://mathworld.wolfram.com/BeastNumber.html
11. Sloane, N. J. A. Sequences A000225, A010785, A048328, A048329, A048330, A048331, A048332, A048333, and A048334 in "The On-Line Encyclopedia of Integer Sequences."
12. Weisstein, E.W. "Bimonster" from *MathWorld.*
13. Conway, J. H. and Sloane, N. J. A. "The Monster Group and its 196884-Dimension Space" and "A Monster Lie Algebra?" Chs. 29-30 in *Sphere Packings, Lattices, and Groups, 2nd ed.* New York: Springer-Verlag, pp. 554-571, 1993.
14. Pegg, E. Jr. and Lomont, C. "Math Games: Evil Numbers." Oct. 4, 2004, www.maa.org/editorial/mathgames/mathgames_10_04_04.html .

Index to Issues 1 through 6 of Recreational Mathematics Magazine: February 1961 through December, 1961

Edited and published by Joseph S. Madachy

This index was originally created by Joseph S. Madachy and published in Issue No. 6, December, 1961 issue of **Recreational Mathematics Magazine**.

Issue and page references are listed as 4:17 meaning issue number 4 (August 1961), page 17. Titles of the individual puzzles, problems, word games, alphametics, geometric dissections are listed under these categories in the subject index. Trivial references to persons or topics are omitted.

Issue numbers correspond to the following issue dates: Issue No. 1 – February 1961; Issue No. 2 – April 1961; Issue No. 3 – June, 1961; Issue No. 4 – August, 1961; Issue No. 5 – October 1961; Issue No. 6 – December 1961.

AUTHOR INDEX

Titles of the author's works are listed alphabetically, an (A) denoting an article. Only the initial appearance of puzzles or problems is given – answer references can be found in the subject index under **Puzzles**.

AMIR-MOEZ, ALI R. Circles and Spirals (A), 5:33; Jamsheed Ghiath-ed-din Kashi (A), 6:49-50; Mathematics of Music (A), 3:31-36

ANDREWS, F. EMERSON Counting by Dozens (A), 3:5-9; Some Sorting Tricks (A), 2:3-5

BAKER, S. **Word Games**: 1:20; 2:7-8; 3:21-22; 4:43-45; 5:45; 6:25-26

BATTERSBY, BERNARD J. Problem in Analysis, 6:47

BAUMGARTNER, WILLIAM 6:46

BEYER, E. A. River-Crossing Dilemma, 4:46

BISSONETTE, DONALD K. Letter to the Editor (Base-n Note), 4:63

BRADBURY, A. G. Alphametic: 4:42, 6:27

BRANSCOME, C. E. Puzzle No. 2, 1:33

BRIGGS, THOMAS S. Analysis of "Square It", 4:52-54

BROOKE, MAXEY Coin-Game Coffee Winners, 6:48. 6:64; Dots and Lines (A), 6:51-55; Haunted Checkerboards (A),3:28-30; Letter to the Editor (Sherlock Holmes), 3:59; Number Curiosities, 1:36, 1:38, 5:53; Pi Paradox, 3:25; Readers' Research Problem, 4:54-55, 5:55; Some Absolutely Amazing Afghan Bands (With J. S. Madachy) (A) 1:47-50; Tessarack, How to Make a Magic (A), 5:40-44; Word Notes 6:26

BROTHER, ALFRED Cross-Number Puzzle, 4:13; Fun, Counting by Sevens (A), 3:10-15; Number System Curiosities with Primes, 5:52; World of Large Numbers (A), 4:28-33

BROWN, ALAN L. Number Curiosities, 3:50, 5:53; Perfect Numbers, 3:54

BUCHMAN, E. O. Readers' Research Analysis (Cube Cutting), 2:38-39

WILLIAMS, MIKE Concentric Circles and Trapezoids, 6:60

TITLE INDEX

The titles of all articles and significant notes are listed alphabetically.

SUBJECT INDEX

Major Departments Are Listed Here

ALPHAMETICS (Answer references are given in parentheses)

PUZZLES (Answer references are given in parentheses)

Readers' Research Problems

WORD GAMES

ERRRATA FOR ISSUES 1 THROUGH 6

Issue -1 February 1961
See Errata: 2:2, 2:16, 2:23, 2:27, 2:47, 2:48

Page 12: Fourth line from bottom, read "West" instead of "East"

Page 13: Add two of Diamonds to East's hand

Page 15: Problem 5 – Delete Six of Hearts and change the Ten of Clubs to the Nine of Clubs

Page 17: Move 13 should read "13. R Mates"

Page 42: Equation 2 should read "-(1)(2)+3! . . . etc."

Page 52: Cube Penetration Table – "5 x 5 x 3   5" should read "5 x 5 x 3   7"

Issue 2 – April 1961

See Errata: 3:2, 3:53

Page 47: Add 605 to the table

Issue 3 – June 1961

See Errata: 3: Cover III, 4:45

Page 37: Answer to RMM SLY GAMES – Change 1226 to 1266

Issue 4 – August 1961

Page 60: In *One-Arm Mathematics* tabulation – change 29738 to 29783

Page 63: Tallman's *Letter* – Seventh line, add "   and a_1, a_k are any of them."

Page 64: Gosling's *Letter* – Last line: The number indicated (which was calculated by the editor) is incorrect. The third palindromic number of the series used in this *Letter* is not yet known.

Issue 5 – October 1961

Page 18: Cross-Number Puzzle – Corrections given on page 63 of the December 1961 issue

Page 25: Pythagorean Magic Square – for 79 read 69

Page 28: 7 x 7 Prime Magic Square – for 111 read 11

Page 30: Figure 5 at Bottom of Page – for EQ in the sixth row read FQ, for EP in the eighth row read FP

Page 32: Figure 10 – the correct diagram is shown below

A	B		E	F		B	A
B	A	E			F	A	B
	E	C	D	D	C	E	
E		D	C	C	D		F
F		D	C	C	D		E
	F	C	D	D	C	E	
B	A	F			E	A	B
A	B		F	E		B	A

Page 37: Mr. Ondrejka's *Letter* – for 34 read 37

Page 60: Answer to the August 1961 Cross-Number Puzzle:
   The sixth line from the top should read 7 0 3 0 1 1 7
   Between the sixth and seventh lines insert: 7 0 5 1 1 6 5

Issue 6 – December 1961
   Page 13: Third line below diagrams, 32b should read 23b
   Page 16: Fifth line from top, for "plane" read "place"

Index to Issues 7 through 12 of Recreational Mathematics Magazine: February 1962 through December, 1962

Edited and published by Joseph S. Madachy

Index prepared by Joseph S. Madachy and published in Issue No. 12, December, 1962 issue of **Recreational Mathematics Magazine**.

Issue and page references are listed as 8:17 meaning issue number (April 1962), page 17. Titles of the individual puzzles, problems, word games, alphametics, geometric dissections are listed under these categories in the subject index. Trivial references to persons or topics are omitted.

Issue numbers correspond to the following issue dates: Issue No. 7 – February 1962; Issue No. 8 – April 1962; Issue No. 9 – June, 1962; Issue No. 10 – August, 1962; Issue No. 11 – October 1962; Issue No. 12 – December 1962.

Author Index: Titles of the authors' works are listed in order of appearance an (A) denoting an article. Only the initial appearance of puzzles or problems is given – answer references can be found in the subject index under **Puzzles** or **Alphametics**.

ABBOTT, ROBERT Babel (A), 8:42-44; ??? (A chessboard game) (A), 10:29-34

AHLBURG, HAYO *Letters to the Editor*: Roulette, 11:17-18

ALLEN, RICHARD K. Four Thousand Years of Easter (A), 11:9-10

AMIR-MOEZ, ALI R. Mathematical Sketches, 8:32, Mathematics and Cards (A), 8:40-41; Aboo-Bakkre Mohammed Al-Karkhi (A), 10:45-46; Ibn Haitham (A), 11:47-48; Ibn Bannaa (A), 12:53-54

ANDERSON, JEAN H., Polyominoes – The "Twenty" Problem (A) 9:25-30

ANONYMOUS What's That Again?, 7:27; Alphametic, 10:11

ASH, AVNER The Magic of One Ninety-Seventh (A), 11:43-45

ASHLEY, G. E. Strange Arithmetic, 8:48; Chinese Arithmetic (A), 12:16-19

BAKER, C. L. Cube Formation Solutions, 9:47

BANKOFF, LEON Cube of an Integer Expressible as the Sum of Three Cubesm 9:46; Number Curiosity about 1962, 11:18

BARR, STEPHEN Cube Formation Again!, 7:24; Two-Animal Puzzle, 7:25; How to Get Into an Argument With a Moebius Stripper (A), 7:28-32

BARTH, LARRY Cartoon, 7:48

BASIN, S. L., The First 571 Fibonacci Numbers (with V. E. Hoggatt, Jr.), 11:19-30

BERG, MURRAY Recent Research in Mersenne Numbers (with Sidney Kravitz), 11:40

BRADBURY, A. G. Alphametics: 7:13, 8:17, 9:19, 10:11, 11:8

BRANSCOME, C. E. Mental Squaring (A), 7:23

BREISCH, RICHARD L. Alphametic, 12:24

BROOKE, MAXEY Fibonacci – Mathematical Innovator (A), 7:42-46

BROTHER U, ALFRED Brain Strainer, 7:27, Primes in Arithmetic Progression (A), 8:50-52; A

Title Index: The titles of all articles and significant notes are listed alphabetically

Subject Index: Major Departments are listed here.

by C. Stanley Ogilvy

PUZZLES (Answer references are given in parentheses)

Errata For Issues 7 Through 12

Issue 7 – February 1962
     Page 31: A '5' should be written in the lower right corner of the bottom figure.
     Pages 37 & 38: Equations 39 & 74 – "$\sqrt{(9!)}$" should read "$(\sqrt{9})!$"
     Page 39: 16th line from bottom – "Palindromic primes" should read "palindromic pairs"
     Page 45: Top equation should read:

$$\frac{a}{b} + \frac{a}{10b^3} + \frac{a}{5b^2}$$

     Page 45: 6th and 4th lines from bottom – interchange 73 and 79
     Page 49: Paragraph (2) – 12749 = (11)(19)(61) and is, therefore, *not* a prime.
     Page 56: 5th line from bottom - "123" should read "128"

Issue 8 – April 1962
     See Errata: 9:52, 10:42
     Page 38: Problem 13 – "3' ½'"" should read "3 1/2'"
     Page 47: Puzzle 3, 7th line from bottom – end of line should read "old as Anne as Anne is in
         years".

Issue 9 – June 1962
     See Errata: 10:42.

Issue 10 – August 1962

Page 5: Last paragraph, first line – "Herodian" should read "Heronian".
Page 13: 5th line: "circle moves" should read "line moves".
Page 35: Problem 5 – The question should have read:
Of how many children was the second grandfather the father?
Page 46: 7th and 16th lines – for "Karhi" read "Karkhi".

Issue 11 – October 1962

Page 11: Line 3 – Tertary should read Tertiary.
2nd paragraph, line 7 – Kabby should read kabby.
2nd paragraph, line 10 – "I add" should be in italics (*I add*)
2nd paragraph, line 11 – *uzhifa* should read *vzhifa*
Second sample sentence (bottom of page) – *de sgifdi* should read *be sgifdi*
Page 12: 1st paragraqh, line 9 – for "opphagite" read "oophagite".
Page 12: 3rd line from bottom – *hhca-dedd* should read *hhac-dedd*.
Page 13: In the Debibi paragraph, line 4 – doxi should read dolxi.
Page 29: The last 50 digits of F_{521} belong to F_{522} and vice versa.

ADDENDA
Issue 10 – August 1962

Page 29: The two free years of RMM go to *Ben R. Ezzell, III* for choosing the name BAROQUE for Mr. Abbott's new game.

Index to Issues 13 through 14 of Recreational Mathematics Magazine: February 1963 through Jan-Feb 1964

Edited and published by Joseph S. Madachy

Index prepared by Charles Ashbacher

Issue and page references are listed as 8:17 meaning issue number (April 1962), page 17. Titles of the individual puzzles, problems, word games, alphametics, geometric dissections are listed under these categories in the subject index. Trivial references to persons or topics are omitted. Issue numbers correspond to the following issue dates: Issue No. 13 – February 1963; Issue No. 14 – Jan-Feb 1964.

Author Index: Titles of the authors' works are listed in order of appearance an (A) denoting an article. Only the initial appearance of puzzles or problems is given – answer references can be found in the subject index under **Puzzles** or **Alphametics**.

AMIR-MOÉZ, ALI R. George Cantor (1845-11918) (A), 13:34-35

BARR, STEPHEN Problem: A Borderline Cover-Uo, 14:43

BERGERSON, HOWARD Problem: An Elementary (?) Problem in Factoring, 14:44

BERGMAN, RONALD (A) Something New Behind the 8-Ball (A), 14:17-19

BICKNELL, MARJORIE (A) (with Verner E. Hoggatt) 64 Ways to Write 64 Using Four 4's, 14:13-15

BRADBURY, A. G. Alphametics, 13:10, 14:46, 13:10, 14:46

BRANSCOME, C. E. Sea Distance (observation), 13:9

BROOKE, MAXEY Sir Isaac Newton, Problemist (A), 13:19; Letter to the Editor (Request for biographical data on Edouard Lucas

BROWN, ALAN L. Multiperfect Numbers – Cousins of the Perfect Numbers (A), 13:30; Multiperfect Numbers – Cousins of the Perfect Numbers – No. 1 (A), 14:31-39

CLID, U. Problem: A Number Problem, 13:39

COHEN, Ben Problem: A Matter of Regions, 14:44

CROSS, D. C. Multigrades (A), 13:7-9; Problem: The King's New Banquet Hall, 13:38

McCLELLAN, JOHN Cartoon, 13:29

MCCLENON, PAUL R. Problem: Scotch and Water, 14:43

MEALLY, VICTOR Letter to the Editor (Dudeney 13 Knights Problem), 13:20

NELSON, HARRY L. Number Curiosity Solution, 13:27-28

PENNEY, WALTER Problem: Many a Mickle, 14:44

PHLPOTT, WADE E. (A) Quadrilles

PROPPER, GEORGE Alphametic, 13:10, 14:46

RANSOM, WILLIAM R. Protean Shapes with Flexagons (A), 13:35-37

RABUCCI, ERNEST R. Junior Department: Non-unique Orthographic Projections, 14:50

READ, RONALD C. Soup, Fish and Finite Geometries (A), 13:11-16; solution 14:46

RUGGLES, D. Book Review: Mathematical Puzzles for Beginners and Enthusiasts, 14:28

SAAR, HOWARD C. Junior Department, 13: 43; Junior Department, 14:49

SANDKUHLE, RAYMOND C. (A) The Equation $A^3 + B^3 = C^3$ 14:15-16

SCOTT, ROBERT H. A Bottle and A Cork, 13:44

SMITH, DAVID Problem: The Bishop Problem, 14:45

SOKOL, LEE Cartoon 13:26

SUTCLIFFE, ALAN Letter to the Editor (Ramanujan Taxi Number), 13:16-17

TESTER, H. E. Problem: Digit Removal, 14:55

THORO, DMITRI E. Book Reviews: 13:23-26; 14:27-30; Book Reviews: Barlow's table of Squares, Cubes, Square Roots, Cube Roots and Reciprocals, 14:29-30; Recreational Mathematics: A Guide to the Literature, 14:30

TRIGG, CHARLES W. Numbers With Distinct Digits of the Form (M-1)M(M+1) (A), 13:27; A Unique Magic Hexagon (A), 14:41-43

UNKNOWN AUTHOR Junior Department Problem Column, Area Doubling, 13:48

VAAGE, EMIL FRIDSTEIN The Elusive Number PI (A), 13:18
VANDERPOOL, DONALD L. Number Curiosity Solution, 13:27

VERBEEK, C. C. Letter to the Editor (Requesting back issues of RMM), 13:20; Problem: Cigarette Selling. 13:38

WALKER, LLOYD A. Letter to the Editor (Contraction symbols used in classroom), 13:22

WEST HIGH SCHOOL MATH CLUB Poetry and Cardiod Curve, 14: 56
WILLIAMS, JEAN M. Book Review: Play Mathematics, 14:28-29

Title Index

64 WAYS TO WRITE 64 USING FOUR 4'S by Marjorie Bicknell and Verner E. Hoggatt 14:13-15

8-BALL, SOMETHING NEW BEHIND THE by Ronald Bergman 14:17-19

ANGLES OF AN OLD TRIANGLE, NEW by Walter W. Horner 14:52-54

BILLIARD BALLS IN AN EQUILATERAL TRIANGLE by Donald E. Knuth 14:20-23

BOTTLE AND A CORK, A by Robert H. Scott 13: 44

CANTOR, GEORGE (1845-1918) by Ali R. Amir-Moéz

CLERIHEW A B C OF MATHEMATICS, A by J. A. Lindon 14:24-26

DECIMALS, RECURRING by Peter Farrell, 13:45-47

EQUATION $A^3 + B^3 = C^3$, THE by Raymond C. Sandkuhle

FLEXAGONS, PROTEAN SHAPES WITH bby William R. Ransom 13:35-37

GEOMETRIC MAGIC SQUARES by Boris Kordemskii, 13:3-6

HEXAGON, A UNIQUE MAGIC by Charles W. Trigg 14:40-43

MAGIC SQUARES, FUN WITH by Dale Kozniuk 14: 50-52

MULTIGRADES by D. C. Cross 13:7-9

MULTIPERFECT NUMBERS – COUSINS OF THE PERFECT NUMBERS, by Alan L. Brown, 13:30

Subject Index

ON A DECONCATENATION PROBLEM

Henry Ibstedt

Glimminge 2036

28060 Broby

Sweden

henry.ibstedt@gmail.com

Abstract

In a recent study of the *Primality of the Smarandache Symmetric Sequences*, Sabin and Tatiana Tabirca [1] observed a very high frequency of the prime factor 333667 in the factorization of the terms of the second order sequence. The question if this prime factor occurs periodically was raised. The odd behaviour of this and a few other primefactors of this sequence will be explained and details of the periodic occurance of this and of several other prime factors will be given.

Definition: The nth term of the Smarandache symmetric sequence of the second order is defined by S(n)=123...n_n...321 which is to be understood as a concatenation of the first n natural numbers concatenated with a concatenation in reverse order of the n first natural numbers.

Note: The underscore character (_) is used to represent concatenation in this paper.

Factorization and Patterns of Divisibility

The first five terms of the sequence are: 11, 1221, 123321, 12344321, 1234554321.

The number of digits D(n) of S(n) grows rapidly. It can be found from the formula:

(1) $$D(n) = 2k(n+1) - \frac{2(10^k - 1)}{9}$$

for n in the interval $10^{k-1} \leq n < 10^k - 1$.

In order to study the repeated occurrance of certain primes S(n) was calculated and partially factorized for n. The results for n ≤ 100 appear in table 1.

The computer file containing table 1 was analysed in various ways. Of the 664579 primes which are smaller than 10^7 only 192 occur in the prime factorizations of S(n) for $1 \leq n \leq 200$. Of these 192 primes 37 occur more than once. The record holder is 333667, the 28693th prime, which occurs 45 times for $1 \leq n \leq 200$ while its neighbours 333647 and 333673 do not occur once. This is summarized in table 2.

Table 1

Prime factors of S(n) which are less than 10^8

n	Prime factors of S(n)	n	Prime factors of S(n)
1	11	51	3 * 37 * 1847 * F180
2	3 * 11 * 37	52	F190
3	3 * 11 * 37 * 101	53	3^3 * 11 * 43 * 26539 * 17341993 * F178
4	11 * 41 * 101 * 271	54	3^3 * 37 * 41 * 151 * 271 * 347 * 463 * 9091 * 333667 * F174
5	3 * 7 * 11 * 13 * 37 * 41 * 271	55	67 * F200
6	3 * 7 * 11 * 13 * 37 * 239 * 4649	56	3 * 11 * F204
7	11 * 73 * 101 * 137 * 239 * 4649	57	3 * 31 * 37 * F206
8	3^2 * 11 * 37 * 73 * 101 * 137 * 333667	58	227 * 9007 * 20903089 * F200
9	3^2 * 11 * 37 * 41 * 271 * 9091 * 333667	59	3 * 41 * 97 * 271 * 9091 * F207
10	F22	60	3 * 37 * 3368803 * F213
11	3 * 43 * 97 * 548687 * F16	61	91719497 * F218
12	3 * 11 * 31 * 37 * 61 * 92869187 * F15	62	3^2 * 1693 * F225
13	109 * 3391 * 3631 * F24	63	3^2 * 37 * 305603 * 333667 * 9136499 * F213
14	3 * 41 * 271 * 9091 * 290971 * F24	64	11 * 41 * 271 * 9091 * F229
15	3 * 37 * 661 * F37	65	3 * 839 * F238

16	F46	66	3 * 37 * 43 * F242
17	3 * F49	67	11^2 * 109 * 467 * 3023 * 4755497 * F233
18	3^2 * 37 * 1301 * 333667 * 6038161 * 87958883 * F28	68	3 * 97 * 5843 * F247
19	41 * 271 * 9091 * F50	69	3 * 37 * 41 * 271 * 787 * 9091 * 716549 * 19208653 * F232
20	3 * 11 * 97 * 128819 * F53	70	F262
21	3 * 37 * 983 * F61	71	3 * F265
22	67 * 773 * F65	72	3^2 * 31 * 37 * 61 * 163 * 333667 * 77696693 * F248
23	3 * 11 * 7691 * F68	73	379 * 323201 * F266
24	3 * 37 * 41 * 43 * 271 * 9091 * 165857 * F61	74	3 * 41^2 * 43^2 * 179 * 271 * 9091 * 8912921 * F255
25	227 * 2287 * 33871 * 611999 * F66	75	3 * 11 * 37 * 443 * F276
26	3^3 * 163 * 5711 * 68432503 * F70	76	1109 * F283
27	3^3 * 31 * 37 * 333667 * 481549 * F74	77	3 * 10034243 * F282
28	146273 * 608521 * F83	78	3 * 11 * 37 * 71 * 41549 * F284
29	3 * 41 * 271 * 9091 * F89	79	41 * 271 * 9091 * F290
30	3 * 37 * 5167 * F96	80	3 * F300
31	11^3 * 4673 * F99	81	3^5 * 37 * 333667 * 4274969 * F289
32	3 * 43 * 1021 * F104	82	F310
33	3 * 37 * 881 * F109	83	3 * 20399 * 5433473 * F302

34	11 * 41 * 271 * 9091 * F109	84	3 * 37² * 41 * 271 * 9091 * F306
35	3² * 3209 * F117	85	1783 * 627041 * F313
36	3² * 37 * 333667 * 68697367 * F110	86	3 * 11 * F324
37	F130	87	3 * 31 * 37 * 43 * F324
38	3 * 1913 * 12007 * 58417 * 597269 * 63800419 * F107	88	67 * 257 * 46229 * F325
39	3 * 37 * 41 * 271 * 347 * 9091 * 23473 * F121	89	3² * 11 * 41 * 271 * 9091 * 653659 * 76310887 * F314
40	F142	90	3² * 37 * 244861 * 333667 * F328
41	3 * 156841 * F140	91	173 * F343
42	3 * 11 * 31 * 37 * 61 * 20070529 * F136	92	3 * F349
43	71 * 5087 * F148	93	3 * 37 * 1637 * F348
44	3² * 41 * 271 * 9091 * 1553479 * F142	94	41 * 271 * 9091 * 10671481 * F343
45	3² * 11 * 37 * 43 * 333667 * F151	95	3 * 43 * 2833 * F356
46	F166	96	3 * 37 * 683 * F361
47	3 * F169	97	11 * 26974499 * F361
48	3 * 37 * 173 * 60373 * F165	98	3² * 1299169 * F367
49	41 * 271 * 929 * 9091 * 34613 * F162	99	3² * 37 * 41 * 271 * 2767 * 9091 * 263273 * 333667 * 481417 * F347
50	3 * 167 * 1789 * 9923 * F172	100	43 * 47 * 53 * 83 * 683 * 3533 * 4919 * F367

Table 2

Counts of the most frequent primes

p	f	p	f
3	132	43	24
33667	45	73	14
37	41	53	8
41	41	97	7
271	41	31	6
9091	29	47	6
11	25		

Obviously there is something deeper happening here. The distribution of the primes 11, 37, 41, 43, 271, 9091 and 333667 is shown in table 3. It is seen that the occurance patterns are different in the intervals $1 \leq n \leq 9$, $10 \leq n \leq 99$ and $100 \leq n \leq 200$. Indeed the last interval is part of the interval $100 \leq n \leq 999$. It would have been very interesting to include part of the interval $1000 \leq n \leq 9999$ but as we can see from (1) already S(1000) has 5784 digits.

Table 3

$p|S(n_0+d\cdot k)$ for k= …

p	n_0	d	k
11	0	1	1,2,…,9
11	9	11	0,1, … ,8
11	12	11	0,1, … ,7
31	12	15	0,1, … ,6
37	2	3	0,5,8
37	3	3	0,2, … ,32
37	99	37	0,2, …, ?

37	122	37	0,2,...,?
41	4	1	0,1
41	9	5	0,1, ... ,?
43	11	21	0,1,2,3 ,4
43	24	21	0,1,2,3
47	100	46	0,1, ... ,?
47	105	46	0,1, ... ,?
53	100	13	0,1, ...?
271	4	1	0,1
271	9	5	0,1, ... ,?
9091	9	5	0,1, ... ,19
9091	99	10	0,1, ... ,?
333667	8	1	0,1
333667	9	9	0,1, ... ,9
333667	99	3	0,1, ...,?

From the patterns in table 3 we can formulate the occurance of these primes in the intervals $1 \leq n \leq 9$, $10 \leq n \leq 99$ and $100 \leq n \leq 200$, where the formulas for the last interval are indicative. We note, for example, that 11 is not a factor of any term in the interval $100 \leq n \leq 999$. This indicates that the divisibility patterns for $1000 \leq n \leq 9999$ and further intervals is a completely open question. There are other primes which also occur periodically but less frequently.

The frequency of the most frequently occurring primes is shown in table 3 in the form $p|S(n_0 + d \cdot k)$, where d is the period and k indicates how far the periodicity is valid. In most cases it is not known – this is indicated by ?. As is seen the periodicity property may or may not change when n passes from 10^α to $10^{\alpha+1}$.

Table 4 shows an analysis of the patterns of occurance of the primes in table 1 by interval. Note that we only have observations up to n = 200. Nevertheless the interval $100 \leq n \leq 999$ is used. This will be justified in the further analysis.

We note that no terms are divisible by 11 for n > 100 in the interval $100 \leq n \leq 200$ and that no term is divisible by 43 in the interval $1 \leq n \leq 9$. Another remarkable observation is that the sequence shows exactly the same behaviour for the primes 41 and 271 in the intervals included in the study. Will they show the same behaviour when n ≥ 1000?

Explanations

Consider $S(n)=12\ldots n\_n\ldots21$.

Let p be a prime divisor of S(n). We will construct a number

(2) $N=12\ldots n\_0..0\_n\ldots21$

so that p also divides N. What will be the number of zeros? Before discussing this let's consider the case p=3.

Table 4
Divisibility patterns

Interval	p	n	Range for j
1≤n≤9	11	All values of n	
10≤n≤99		12+11j	j=0,1, … ,7
		20+11j	j=0,1, … ,7
100≤n≤999		None	
1≤n≤9	37	2+3j	j=0,1,2
		3+3j	j=0,1,2
10≤n≤99		12+3j	j=0,1,…,28,29
100≤n≤999		122+37j	j=0,1,…,23
		136+37j	j=0,1,…,23

$1 \leq n \leq 9$	41	$4+5j$	$j=0,1$
		5	
$10 \leq n \leq 999$		$14+5j$	$j=0,1,\ldots,197$
$1 \leq n \leq 9$	43	None	
$10 \leq n \leq 99$		$11+21j$	$j=0,1,3,4$
		$24+21j$	$j=0,1,2,3$
$100 \leq n \leq 999$		100	
		$107+7j$	$j=0,1,\ldots,127$
$1 \leq n \leq 9$	271	$4+5j$	$j=0,1$
		5	
$10 \leq n \leq 999$		$14+5j$	$j=0,1,\ldots,197$
$1 \leq n \leq 999$	9091	$9+5j$	$j=0,1,\ldots,98$
$1 \leq n \leq 9$	333667	8,9	
$10 \leq n \leq 99$		$18+9j$	$j=0,1,\ldots,9$
$100 \leq n \leq 999$		$102+3j$	$j=0,1,\ldots,299$

Case 1: p = 3.

In the case p = 3 we use the familiar rule that a number is divisible by 3 if and only if its digit sum is divisible by 3. In this case we can insert as many zeros as we like in (2) since this does not change the sum of the digits. We also note that any integer formed by concatenation of three consecutive integers is divisible by 3, cf a_a+1_a+2, has digit sum 3a+3.

It follows that also a_a+1_a+2_a+2_a+1_a is divisible by 3. For a = n + 1 we insert this instead of the appropriate number of zeros in (2). This means that if S(n) ≡ 0 (mod 3) then S(n+3) ≡ 0 (mod 3). We have seen that S(2) ≡ 0 (mod 3) and S(3) ≡ 0 (mod 3). By induction it follows that

S(2+3j) ≡ 0 (mod 3) for j = 1,2,… and S(3j) ≡ 0 (mod 3) for j=1,2,… .

We now return to the general case. S(n) is deconcatenated into two numbers 12…n and n… 21 from which we form the numbers

$$A = 12...n \cdot 10^{1+[\log_{10} B]}$$

and B = n...21.

We note that this is a different way of writing S(n) since indeed A + B = S(n) and that
A + B ≡ 0 (mod p). We now form

M = A·10^s + B where we want to determine s so that

M ≡ 0 (mod p). We write M in the form M = A(10^s - 1) + A + B where A + B can be ignored
mod p. We exclude the possibility

A ≡ 0 (mod p) which is not interesting. This leaves us with the congruence

$$M \equiv A(10^s-1) \equiv 0 \ (\text{mod } p)$$

or

$$10^s - 1 \equiv 0 \ (\text{mod } p).$$

We are particularly interested in solutions for which

pε { 11, 37, 41, 43, 271, 9091, 333667 }.

By the nature of the problem these solutions are periodic. Only the first two values of s are given
for each prime.

Table 5

10^s-1≡0 (mod p)

p	3	11	37	41	43	271	9091	333667
s	1,2	2,4	3,6	5,10	21,42	5,10	10,20	9,18

We note that the result is independent of n. This means that we can use n as a parameter when
searching for a sequence

$$C=n+1\_n+2\_...n+k\_n+k\_...n+2\_n+1$$

such that this is also divisible by p and hence can be inserted in place of the zeros to form
S(n + k) which then fills the condition S(n+k) ≡ 0 (mod p). Here k is a multiple of s or s / 2 in
case s is even. This explains the results which we have already obtained in a different way as

89

part of the factorization of S(n) for n ≤ 200, see tables 3 and 4. It remains to explain the periodicity which as we have seen is different in different intervals $10^u \le n \le 10^u - 1$.

This may best be done by using concrete examples. Let us use the sequences starting with n = 12 for p = 37, n = 12 and n = 20 for p = 11 and n = 102 for p = 333667. At the same time we will illustrate what we have done above.

Case 2: n = 12, p = 37. Period = 3. Interval: 10 ≤ n ≤ 99.

S(n) = 123456789101112_____121110987654321

N = 12345678910111200000000000000121110987654321

C = 131415151413

S(n+k) = 123456789101112131415151413121110987654321.

Let's look at C which carries the explanation to the periodicity.
We write C in the form

$$C = 101010101010 + 30405050403.$$

We know that C ≡ 0 (mod 37). What about 101010101010? Let's write

$$101010101010 = 10 + 10^3 + 10^5 + \ldots + 10^{11} =$$
$$(10^{12} - 1) / 9 \equiv 0 \ (\text{mod } 37).$$

This congruence mod 37 has already been established in table 5.

It also follows that

$$30405050403 \equiv 0 \ (\text{mod } 37)$$

and that
$$x * (101010101010) \equiv 0 \ (\text{mod } 37) \ \text{for } x = \text{any integer}.$$

Combining these observations we see that

232425252423, 333435353433, … 939495959493 ≡ 0 (mod 37).

Hence the periodicity is explained.

Case 3a: n = 12, p = 11. Period = 11. Interval: $10 \leq n \leq 99$.

S(12) = 12_.._12   12_.._21.

S(23) = 12_.._1213141516171819202122232322212019181716
15141312_.._21.

C = 13141516171819202122232322212019181716151413 =

C1 = 1010101010101010101010101010101010101010101010 +

C2 = 3040506070809101112131312111009080706050403.

From this we form

2 * C1 + C2 = 2324252627282930313233333231302928272625242

which is NOT what we wanted, but $C1 \equiv 0 \ (mod 11)$ and also
$C1 \ / \ 10 \equiv 0 \ (mod \ 11)$. Hence we form

2 * C1 + C1 / 10 + C2 = 2425262728293031323334343332313029282726252

which is exactly the C-term required to form the next term S(34) of the sequence. For the next
term S(45) the C-term is formed by 3 * C1 + 2 * C1 / 10 + C2. The process is repeated adding
C1 + C1 / 10 to proceed from a C-term to the next until the last term < 100, i.e. S(89) is reached.

Case 3b: n = 20, p = 11. Period = 11. Interval: $10 \leq n \leq 99$.

This case does not differ much from the case n = 12. We have

S(20) = 12_.._20   20_.._21

S(31) = 12_.._20212223242526272829303131302928272625 24
23222120_.._21
C = 212223242526272829303131302928272625242322 21 =

C1 = 1010101010101010101010101010101010101010101010 +

C2 = 102030405060708091011111009080706050403020 1.

The C-term for S(42) is

$3 * C1 + C1 / 10 + C2 = 32333435363738394041424241403938373635343332.$

In general $C = x * C1 + (x - 1) * C1 / 10 + C2$ for $x = 3,4,5, ..,8$.

For $x = 8$ the last term $S(97)$ of this sequence is reached.

Case 4: $n = 102$, $p = 333667$. Period = 3. Interval: $100 \leq n \leq 999$.

$S(102)=12\_..\_101102_____102101\_..\_21$

$S(105)=12\_..\_101102103104105105104103102101\_..\_21$

$C = 103104105105104103 \equiv 0 \pmod{333667}$
$C1 = 100100100100100100 \equiv 0 \pmod{333667}$
$C2 = 3004005005004003 \equiv 0 \pmod{333667}$.

Removing 1 or 2 zeros at the end of C1 does not affect the congruence modulus 333667, we have:

$C1' = 10010010010010010 \equiv 0 \pmod{333667}$
$C1'' = 1001001001001001 \equiv 0 \pmod{333667}$.

We now form the combinations:

$x * C1 + y * C1' + z * C1'' + C2 \equiv 0 \pmod{333667}$.

This, in my mind, is quite remarkable. All 18-digit integers formed by the concatenation of three consecutive 3-digit integers followed by a concatenation of the same integers in descending order are divisible by 333667, for example $376377378378377376 \equiv 0 \pmod{333667}$. As far as the C-terms are concerned all $S(n)$ in the range $100 \leq n \leq 999$ could be divisible by 333667, but they are not. Why? It is because $S(100)$ and $S(101)$ are not divisible by 333667. Consequently $n =100 + 3k$ and $101 + 3k$ cannot be used for insertion of an appropriate C-value as we did in the case of $S(102)$. This completes the explanation of the remarkable fact that every third term $S(102 + 3j)$ in the range $100 \leq n \leq 999$ is divisible by 333667.

These three cases have shown what causes the periodicity of the divisibility of the Smarandache symmetric sequence of the second order by primes. The mechanism is the same for the other periodic sequences.

Beyond 1000

We have seen that numbers of the type:

10101010...10, 100100100...100, 10001000...1000,

etc. play an important role. Such numbers have been factored and the occurrence of our favorite primes 11, 37, ..., 333667 have been listed in table 6. In this table a number like 100100100100 has been abbreviated 4(100) or q(E), where q and E are listed in separate columns.

Table 6

Prime factors of q(E) and occurrence of selected primes

q	E	Prime factors <350000	Selected primes
2	10	2 * 5 * 101	
3		2 * 3 * 5 * 7 * 13 * 37	37
4		2 * 5 * 73 * 101 * 137	
5		2 * 5 * 41 * 271 * 9091	41,271,9091
6		2 * 3 * 5 * 7 * 13 * 37 * 101 * 9901	37,9091
7		2 * 5 * 239 * 4649 *	
8		2 * 5 * 17 * 73 * 101 * 137 *	
9		2 * 3^2 * 5 * 7 * 13 * 19 * 37 * 52579 * 333667	333667
10		2 * 5 * 41 * 101 * 271 * 3541 * 9091 * 27961	41,271,9091
11		2 * 5 * 11 * 23 * 4093 * 8779 * 21649 *	11
12		2 * 3 * 5 * 7 * 13 * 37 * 73 * 101 * 137 * 9901 *	37
13		2 * 5 * 53 * 79 * 859 *	
14		2 * 5 * 29 * 101 * 239 * 281 * 4649 *	

15		2 * 3 * 5 * 7 * 13 * 31 * 37 * 41 * 211 * 241 * 271 * 2161 * 9091 *	37,41,271,9091
16		2 * 5 * 17 * 73 * 101 * 137 * 353 * 449 * 641 * 1409 * 69857 *	
2	10^2	2^2 * 5^2 * 7 * 11 * 13	11
3		2^2 * 3 * 5^2 * 333667	333667
4		2^2 * 5^2 * 7 * 11 * 13 * 101 * 9901	11
5		2^2 * 5^2 * 31 * 41 * 271 *	41,271
6		2^2 * 3 * 5^2 * 7 * 11 * 13 * 19 * 52579 * 333667	11,333667
7		2^2 * 5^2 * 43 * 239 * 1933 * 4649 *	43
8		2^2 * 5^2 * 7 * 11 * 13 * 73 * 101 * 137 * 9901 *	11,73
9		2^2 * 3^2 * 5^2 * 757 * 333667 *	333667
10		2^2 * 5^2 * 7 * 11 * 13 * 31 * 41 * 211 * 241 * 271 * 2161 * 9091 *	11,41,271,9091
11		2^2 * 5^2 * 67 * 21649 *	
12		2^2 * 3 * 5^2 * 7 * 11 * 13 * 19 * 101 * 9901 * 52579 * 333667 *	11,333667
2	10^3	2^3 * 5^3 * 73 * 137	
3		2^3 * 3 * 5^3 * 7 * 13 * 37 * 9901	37
4		2^3 * 5^3 * 17 * 73 * 137 *	
5		2^3 * 5^3 * 41 * 271 * 3541 * 9091 * 27961	41,271,9091
6		2^3 * 3 * 5^3 * 7 * 13 * 37 * 73 * 137 * 9901 *	37
7		2^3 * 5^3 * 29 * 239 * 281 * 4649 *	
8		2^3 * 5^3 * 17 * 73 * 137 * 353 * 449 * 641 * 1409 * 69857 *	
9		2^3 * 3^2 * 5^3 * 7 * 13 * 19 * 37 * 9901 * 52579 * 333667 *	37,333667

10		2^3 * 3 * 5^3 * 41 * 73 * 137 * 271 * 3541 * 9091 * 27961 *	41,271,9091
11		2^3 * 5^3 * 11 * 23 * 89 * 4093 * 8779 * 21649 *	11
2	10^4	2^4 * 5^4 * 11 * 9091	11,9091
3		2^4 * 3 * 5^4 * 31 * 37 *	37
4		2^4 * 5^4 * 11 * 101 * 3541 * 9091 * 27961	11,9091
5		2^4 * 5^4 * 21401 * 25601 *	
6		2^4 * 3 * 5^4 * 7 * 11 * 13 * 31 * 37 * 211 * 241 * 2161 * 9091 *	11,37,9091
7		2^4 * 5^4 * 71 * 239 * 4649 * 123551 *	
8		2^4 * 5^4 * 11 * 73 * 101 * 137 * 3541 * 9091 * 27961 *	11,9091
9		2^4 * 3 * 5^4 * 31 * 37 * 238681 * 333667 *	37,333667
2	10^5	2^5 * 5^5 * 101 * 9901	
3		2^5 * 3 * 5^5 * 19 * 52579 * 333667	333667
4		2^5 * 5^5 * 73 * 101 * 137 * 9901 * *	
5		2^5 * 5^5 * 31 * 41 * 211 * 241 * 271 * 2161 * 9091 * *	41,271,9091
6		2^5 * 3 * 5^5 * 19 * 101 * 9901 * 52579 * 333667 * *	333667
7		2^5 * 5^5 * 7 * 43 * 127 * 239 * 1933 * 2689 * 4649 * *	43
8		2^5 * 5^5 * 17 * 73 * 101 * 137 * 9901 * *	
9		2^5 * 3^2 * 5^5 * 19 * 757 * 52579 * 333667 * *	333667

Question 1: Does the sequence of terms S(n) divisible by 333667 continue beyond 1000?

Although S(n) was partially factored only up to n = 200 we have been able to draw conclusions on divisibility up to n = 1000. The last term that we have found divisible by 333667 is S(999). Two conditions must be met for there to be a sequence of terms divisible by p = 333667 in the interval 1000 ≤ n ≤ 9999.

Condition 1. There must exist a number 10001000...1000 divisible by 333667 to ensure the periodicity as we have seen in our case studies.

In table 7 we find q = 9, E = 1000. This means that the periodicity will be 9 – if it exists, i.e. condition 1 is met.

Condition 2. There must exist a term S(n) with n ≥ 1000 divisible by 333667 which will constitute the first term of the sequence.

The last term for n < 1000 which is divisible by 333667 is S(999) from which we build

S(108)=12...999_1000_..._1008_1008_...1000_999-...21

where we deconcatenate 100010011002...10081008...10011000 which is divisible by 333667 and provides the C-term (as introduced in the case studies) needed to generate the sequence, i.e. condition 2 is met.

We conclude that S(1008+9j) ≡ 0 (mod 333667) for j = 0,1,2, ..., 999. The last term in this sequence is S(9999). From table 7 we see that there could be a sequence with the period 9 in the interval 10000 ≤ n ≤ 99999 and a sequence with period 3 in the interval 100000 ≤ n ≤ 999999. It is not difficult to verify that the above conditions are also satisfied in these intervals. This means that we have:

S(1008 + 9j) ≡ 0 (mod 333667) for j = 0, 1,2,...,999, i.e.
$10^3 \leq n \leq 10^4 - 1$.

S(10008 + 9j) ≡ 0 (mod 333667) for j = 0, 1,2,...,9999, i.e.
$10^4 \leq n \leq 10^5 - 1$

S(100002 + 3j)≡0 (mod 333667) for j = 0, 1,2,...,99999, i.e.
$10^5 \leq n \leq 10^6 - 1$.

It is one of the fascinations with large numbers that you can find such properties. This extraordinary property of the prime 333667 in relation to the Smarandache symmetric sequence probably holds for n > 10⁶. It easy to loose contact with reality when playing with numbers like this. We have S(999999) ≡ 0 (mod 333667).

What does this number S(999999) look like? Applying (1) we find that the number of digits D(999999) of S(999999) is

$$D(999999) = 2 * 6 * 10^6 - 2 * (10^6) / 9 = 11777778.$$

Let's write this number with 80 digits per line, 60 lines per page, using both sides of the paper. We will need 1226 sheets of paper – more that 2 reams!

Question 2. Why is there no sequence of S(n) divisible by 11 in the interval $100 \leq n \leq 999$?

Condition1. We must have a sequence of the form 100100.. divisible by 11 to ensure the periodicity. As we can see from table 7 the sequence 100100 fills the condition and we would have a periodicity equal to 2 if the next condition is met.

Condition 2. There must exist a term S(n) with n ≥ 100 divisible by 11 which would constitute the first term of the sequence. This time let's use a nice property of the prime 11:

$$10^s \equiv (-1)^s \pmod{11}.$$

Let's deconcatenate the number a_b corresponding to the concatenation of the numbers a and b: We have:

$$a\_b = a \cdot 10^{1+[\log_{10} b]} + b = \begin{cases} -a+b & \text{if } 1+[\log_{10} b] \text{ is odd} \\ a+b & \text{if } 1+[\log_{10} b] \text{ is even} \end{cases}$$

Let's first consider a deconcatenated middle part of S(n) where the concatenation is done with three-digit integers. For convienience I have chosen a concrete example – the generalization should pose no problem

$$273274275275274273 \equiv$$
$$2 - 7 + 3 - 2 + 7 - 4 + 2 - 7 + 5 - 2 + 7 - 5 + 2 - 7 + 4 - 2 + 7 - 3 \equiv$$
$$0 \pmod{11}$$
$$+ - + - + - + - + - + - + - + - + -.$$

It is easy to see that this property holds independent of the length of the sequence above and whether it starts on a + or a -. It is also easy to understand that equivalent results are obtained for other primes although factors other than +1 and −1 will enter into the picture.

97

We now return to the question of finding the first term of the sequence. We must start from $n = 97$ since $S(97)$ it the last term for which we know that $S(n) \equiv 0 \pmod{11}$. We form:

$9899100101...n\_n...1011009998 \equiv 2 \pmod{11}$ independent of
$n < 1000$.

+-+-+-+-+-... _ ...-+-+-+-+-

This means that $S(n) \equiv 2 \pmod{11}$ for $100 \le n \le 999$ and explains why there is no sequence divisible by 11 in this interval.

Question 3. Will there be a sequence divisible by 11 in the interval $1000 \le n \le 9999$?

Condition 1. A sequence $10001000...1000$ divisible by 11 exists and would provide a period of 11, see table 6.

Condition 2. We need to find one value $n \ge 1000$ for which $S(n) \equiv 0 \pmod{11}$. We have seen that $S(999) \equiv 2 \pmod{11}$. We now look at the sequences following $S(999)$. Since $S(999) \equiv 2 \pmod 9$ we need to insert a sequence

$10001001..m\_m...10011000 \equiv 9 \pmod{11}$

so that $S(m) \equiv 0 \pmod{11}$. Unfortunately m does not exist as we will see below.

$10001000 \equiv 2 \pmod{11}$
+-+-+-+-
1   1
$1000100110011000 \equiv 2 \pmod{11}$
+-+-+-+-+-+-+-
1 1 1 1
    1 1
$100010011002100210011000 \equiv 0 \pmod{11}$
+-+-+-+-+-+-+-+-+-+-
1 1 1 1 1 1
    1 2 2 1
$1000100110021003100310 02100210011000 \equiv -4 \equiv 7 \pmod{11}$
+-+-+-+-+-+-+-+-+-+-+-+-+
1 1 1 1 1 1 1 1
    1 2 3 3 2 1.

Continuing in this manner we find that the residues form the period 2, 2, 0, 7, 1, 4, 5, 4, 1, 7, 0. We needed a residue to be 9 in order to build sequences divisible by 9. We conclude that S(n) is not divisible by 11 in the interval $1000 \leq n \leq 9999$.

Trying to do the above analysis with the computer programs used in the early part of this study causes an overflow due to the large integers involved. However, changing the approach and performing calculations modulo 11 posed no problems. The above method was preferred for clarity of presentation.

Epilog

There are many other questions that may be interesting to look into. This is left to the reader. The author's main interest in this has been to develop means by which it is possible to identify some properties of large numbers other than the frequently asked question as to whether a big number is a prime or not. There are two important ways to generate large numbers which I found particularly interesting – iteration and concatenation. In this article the author has drawn on work done previously, see the references that follow. In both these areas very large numbers may be generated for which it may be impossible to find any practical use – the methods are often more important than the results.

References

1. Tabirca, S. and T., *On Primality of the Smarandache Symmetic Sequences,* Smarandache Notions Journal, Vol. 12, No 1-3 Spring 2001, 114-121.
2. Smarandache F., *Only Problems, Not Solutions*, Xiquan Publ., Pheonix-Chicago, 1993.
3. Ibstedt H. *Surfing on the Ocean of Numbers,* Erhus University Press, Vail, 1997.
4. Ibstedt H, *Some Sequences of Large Integers,* Fibonacci Quarterly, 28(1990), 200-203.

ALPHAMETICS

Edited by Charles Ashbacher

All of the alphametics in this section were created by Charles Ashbacher

1. This is a doubly true alphametic in Western Apache

```
    dałaa        1
    dałaa        1
    dałaa        1
    taàgi        3
_____    _____
    gostàn       6
```

2. This is a doubly true alphametic in Munsee Delaware. Since the problem as stated has multiple solutions, we add the restriction that the sum should be minimal.

```
    NGWUT        1
    NGWUT        1
    NGWUT        1
    NGWUT        1
    NGWUT        1
    NGWUT        1
    NGWUT        1
    NGWUT        1
    NGWUT        1
_____    _____
    NOOLII       9
```

3. Doubly true Pima Bajo. This is a language of a Native American tribe that lives in the mountainous region on the border between the states of Chihuahua and Sonora in northern Mexico.

```
    HɨMAK        1
    HɨMAK        1        Since the solution is not unique, we will minimize GIGIKO
    HɨMAK        1
    HɨMAK        1
    HɨMAK        1
    HɨMAK        1
_____    _____
    GIGIKO       6
```

4. This problem is doubly true in the Pame language, a language of a Native American tribe in Mexico. At this time there are approximately 10,000 native speakers

```
  NDA        1
  NDA        1     Since there are multiple solutions, we minimize TERIA
RANHU        3
_____   ____

TERIA        5
```

5. This problem is doubly true in the Native American language of the Atakapa tribe. They lived in southwestern Louisiana and east Texas and the language went extinct in the early 20th century.

```
  HAPAL      2
  HAPAL      2     Since there are multiple solutions, we minimize WOCPE
    LAT      3
    LAT      3
_____   ____

  WOCPE     10
```

6. This problem is doubly true in Yaqui, the language of a Native American tribe in the state of Sonora in Mexico and in the southwestern United States. Their language and culture still exists.

```
  SEENU      1
  SEENU      1     Since the solution is not unique, we minimize VATANI
   VAHI      3
  MAMNI      5
_____   ____

 VATANI     10
```

7. This problem is doubly true in Zapotec, a language spoken by Native Americans in the southwestern-central highlands of Mexico. Their language and culture still exists.

```
  TOBI       1
  TOBI       1     Since the solution is not unique, we minimize GADXE
  TOBI       1
  TAPA       4
_____   ____

 GADXE       7
```

8. This problem is doubly true in Tonkawa, a language spoken by Native American tribes indigenous to Oklahoma and Texas. The language went extinct in the twentieth century and it is now considered to be a language isolate in that is not related to any other language.

```
  KETAY      2
  METIS      3
  KETAY      2
  METIS      3
  _____   ____
  SIKPAX    10
```

9. In the "Star Trek: The Next Generation" episode "Chain of Command Part II" the captive Captain Picard yells at his Cardasian torturer "There are four lights!" It was episode number 137 of the series and this alphametic is based on those facts. Since it was a moment of emphasis in the series, we consider the exclamation point to be part of the puzzle and maximize the value of FOUR!

```
     137
   THERE
     ARE
   FOUR!
  _____

  LIGHTS       .
```

BOOK REVIEWS

Edited by:Charles Ashbacher

Charles Ashbacher Technologies

5530 Kacena Ave

Marion, IA 52302

E-mail: cashbacher@yahoo.com

The Lego Adventure Book, Volume 3: Robots, Planes, Cities & More! , by Megan Rothrock, No Starch Press, San Francisco, California, 2015. 192 pp., $24.95(hardback). ISBN 9781593276102.

This book is a testament to the incredible creativity of humans, in this case it is expressed in the construction of very elaborate structures made from LEGO pieces. Several artists are featured, they are from several countries and what they have built are famous structures in their countries. For example, what may be the most elaborate is a duplicate of Groothoofdspoort: the Dordrecht City Gate. The artist's name is Patrick Bosman and he is a Dutch historic preservation advisor.

Detailed, colored plans for each of the structures are presented as a set of sequential steps, moving from caption to caption. While it may take the purchase of several Lego kits and a great deal of patience, if you follow the instructions it is possible to build the beautiful structures seen on these pages.

I was amazed at what these people built, their skill at taking an image of a physical structure and using LEGO bricks to create a duplicate is astonishing. The sequential appearance of the buildings and scenes is punctuated by a running dialog of the cartoon "good guys" engaged in a battle with the forces of destruction, called the destructors.

The scenes even include planes, cars and other auxiliary elements that round out the scenes and make them look even more lifelike. If you love LEGOS and what you can do with them, this is a book that you must read and enjoy. Links to additional images of their creations are included as well. After looking through this book you will understand the legitimacy of the phrase "LEGO artist."

Charles Ashbacher

Computation, Proof, Machine: Mathematics Enters A New Age, by Gilles Dowek, Cambridge University Press, New York, New York, 2015. 152 pp., $39.99(paperback). ISBN 978-0-521-13377-7.

The idea of resolving mathematical uncertainties via computation has been in mathematics for centuries. The most famous original expression of that idea was by Gottfried Wilhelm Leibniz in the seventeenth century when he was describing disputes among persons and he said, "... we can simply say: Let us calculate, without further ado, to see who is right."

Adding machines and computers were used from the moment of their creation to perform numeric computations and resolve some outstanding issues, but the machine did nothing but perform glorified arithmetic. Since the operations were arithmetic, verification was tedious, but still possible.

That changed in 1976 when Kenneth Appel and Wolfgang Haken announced their proof of the four color theorem. The proof was revolutionary in that the rigor of the proof was provided by a computer program that evaluated nearly 2,000 different map structures to reach the conclusion. This was the first major mathematical result done by computers where there was no human verification. The announcement literally changed the definition of the phrase "rigorous proof."

Dowek reviews the history of computing within the context of mathematics, how the art of computation is changing mathematics and how more and more of mathematical progress is defined by the improvement in and development of new computational algorithms. The author begins with mathematical prehistory and ends with Turing machines, Church's thesis and the growing length and complexity of proofs.

This is a book that could be used in a special topics class in both mathematics and computer science. It is a look at the past as well as some very logical speculations as to how both fields will simultaneously advance in the future. There are also some major philosophical points. For example, in chapter fourteen, "The End of Axioms?" Dowek discusses the idea whether computation rules will replace traditional axioms.

Charles Ashbacher

The Art of Lego Scale Modeling, by Dennis Glaasker and Dennis Boseman, No Starch Press, San Francisco, California, 2015. 204 pp, $29.95(hardbound). ISBN 978-1-59327-615-7.

This book contains some of the most amazing creations that I have ever seen, scale models of large and small vehicles made using LEGO blocks and accessories. For example, on pages 22-25 there is a tractor truck car carrier with several cars of different models on the racks.

The item categories in the book are:

*) Trucks

*) Ships

*) Aircraft

*) High performance racing cars

*) Heavy equipment

*) Trains

*) Military fighting vehicles

*) Motorcycles

*) Cars

While some of the models required specialized LEGO blocks, that does not detract from the incredible skill and imagination needed to build them. There are times when you have to look very carefully to see that the structure was in fact build using LEGOS. I came away very impressed with what the people featured in this book accomplished.

Charles Ashbacher

Trigonometry: A Clever Study Guide, by James Tanton, the Mathematical Association of America, Washington, D. C., 2015. 211 pp., $19.95 (paper). ISBN 978-0883858363.

This is as much of a guide to solving problems as it is to the study of trigonometry. Tanton opens each chapter with an explicit statement of the Common Core topics that are being addressed in that chapter. They are a reminder that in mathematics, a common core of learned topics has been a part of our agenda for a long time. Using text and diagrams, the topics of the chapter are then explained.

This is followed by a featured problem for the chapter, where the author recapitulates the process that he followed in solving it. These problems are not bunnies, they require some thought and insight when solving them. The last part of the chapter is a set of problems that have appeared in mathematical contests and solutions to all of them appear in the second section. They are multiple choice problems.

The real unique feature of this book is the set of ten problem solving strategies that Tanton employs when he solved the featured problems. They are:

*) Engage in successful flailing – this means simply writing down whatever it is you know about the problem.

*) Do something – just the act of writing down something related but not necessarily relevant can trigger a valuable thought.

*) Engage in wishful thinking – in this strategy if you need something, write it down. For example, if you need a +4 on one side of an equation put it on both sides and see what you have.

*) Draw a picture – always a sound strategy in mathematics, when possible

*) Solve a smaller version of the same problem – or equivalently, solve a part of it if the problem can be segmented

*) Eliminate incorrect choices – a standard tactic on a multiple choice question

*) Perseverance is key – in other words, keep trying to examine the problem, look at it different ways

*) Second-guess the author – look for clues in the statement, for example the mention that 131 is prime. Irrelevant or a key clue?

*) Avoid hard work – specifically working out the precise value of large numbers. Is there a pattern at work?

*) Go to extremes – for example in a puzzle that deals with the ages of people, what if they were all the same age? What if the escalator was not moving?

Tanton references these strategies in his descriptions of solving the featured problems and manages to inject a bit of humor into the work.

Trigonometry is a topic that is core to the understanding of basic mathematics and in this book Tanton emphasizes that even experienced mathematicians will look at problems and have an initial reaction of "What?" If you teach trig, there are many interesting and effective pedagogical techniques in this book as well as problems that you can use.

Charles Ashbacher

A Mathematical Space Odyssey: Solid Geometry in the 21st Century, by Claudi Alsina and Roger B. Nelsen, The Mathematical Association of America, Washington, D. C., 2015. 272 pp., $55.00 (hardbound). ISBN 978-0-88385-358-0.

It is unfortunate that the teaching of solid geometry has largely ceased in American high schools and colleges. It would seem to be a natural topic, for it is one of the areas of mathematics where the student can be exposed to understandable diagrams that reinforce the subject matter. The authors lament this loss and by writing this book, are doing something about it.

Although calculus is occasionally used when it is necessary to make the appropriate point, in general a knowledge of high school algebra is all the background needed. The one critical skill is an ability to think and maneuver in three dimensions. A large number of three-dimensional figures are described and analyzed, the reader needs to be able to conceptualize the shapes as well as what it means to pass lines and planes through them.

This is a text where a set of companion manipulatives would make for a class that the mathematically balanced art student could take. Specifically the art student interested in sculpture or architecture, there are several images that reference structures and other practical applications. There have been many times when a student in a math class has come to me with difficulties and has said, "I am a visual learner."

Geometry is almost certainly the oldest area of abstract mathematics, for it grew out of the most practical of operations, interacting with and manipulating the physical world. Solid geometry and the rules defining and describing the objects are used in many areas, from machining custom parts to the construction of massive buildings. This book provides a valuable resource for the study of solid geometry that could be used by teachers all the way from late elementary school through college.

A set of challenge problems are given at the end of each chapter and solutions to all appear at the end.

Charles Ashbacher

Solutions to Alphametics in This Issue

Charles Ashbacher

1.

```
    42622
    42622
    42622
    52713
   _____

   180579
```

2.

```
    18453
    18453
    18453
    18453
    18453
    18453
    18453
    18453
    18453
   _____

   166077
```

3.

```
    69023
    69023
    69023
    69023
    69023
    69023
   _____

   414138
```

4.

```
    798
    798
  38762
  ───────
  40358
```

5.

```
  13435
  13435
    538
    538
  ───────
  27946
```

6.

```
  37746
  37746
   1208
  52548
  ───────
  129248
```

7.

```
  2758
  2758
  2758
  2090
  ───────
  10364
```

8.

```
  43701
  83752
  43701
  83752
  ───────
  254906
```

9.

			1	3	7
	10	0	3	2	3
			4	2	3
	13	11	6	2	8
1	9	12	0	10	7

Problems And Conjectures

Edited by: Lamarr Widmer

Readers are invited to send new problems, solutions and comments to me at *Messiah College, Suite 3041, One College Avenue, Mechanicsburg, PA 17055* or email to widmer@messiah.edu . Put each problem or solution, with your full name and postal address, on a separate sheet. Selection of solutions for publication will take place at least three months after problems appear in print.

1. Polyhedron Puzzle by Brian Barwell, Hampton, Middlesex, UK

A polyhedron has faces of three types: squares, regular hexagons and regular decagons. At each of its vertices three faces meet, one of each type. How many faces does the polyhedron have?

2. Altered Prime Generates Additional Primes by Hubert Hagadorn, Menlo Park, CA

A designated digit in a prime number p may be altered to create additional primes. What is the maximum number of primes which may be produced by this alteration procedure? What is the smallest p which allows this maximum number of new primes?

3. Un-powered Numbers by Hubert Hagadorn, Menlo Park, CA

Knowledge of the last two digits of an integer may be sufficient to conclude that it is not an integral power of any smaller integer. For what pairs of final digits is this the case?

4. A Triangular Inequality by Henry Ibstedt, Issy les Moulineaux, France

Prove that the following inequality holds in any triangle ABC.

$$\frac{1}{1 + \sin\frac{A}{2}} + \frac{1}{1 + \sin\frac{B}{2}} + \frac{1}{1 + \sin\frac{C}{2}} \geq 2$$

5. A Hard Way to Tell Time by Hubert Hagadorn, Menlo Park, CA

On a standard 12-hour clock, the hour and minute hands move from an initial condition where they both are between the same numbers on the clock face. They move until the sum of the numbers passed by both hands is 101. What time is it when this sum is attained?

Solutions To Problems From TRM Volume 1

Contributed by Lamarr Widmer

Messiah College
Suite 3041, One College Avenue
Mechanicsburg, PA 17055

1. Infinite Sequence of Integral Square Roots by Henry Ibstedt, Issy les Moulineaux, France

Prove that, for a suitably chosen integer x_1, the recursive formula

$$x_{n+1} = x_n + 4 + 4\sqrt{x_n - a}$$

where a is a positive integer, produces an infinite sequence of integers.

Solution by Kathleen Lewis

Let $x_1 = a$. We will show that $x_n = (2n - 2)^2 + a$, for all $n \in \mathbb{N}$. This is clearly true for $n = 1$. Assuming it true for $n = k$, we have

$$x_{k+1} = x_k + 4 + 4\sqrt{x_k - a} = (2k - 2)^2 + a + 4 + 4\sqrt{(2k - 2)^2 + a - a} = (2k - 2)^2 + 4(2k - 2) + 4 + a = (2k)^2 + a$$

By mathematical induction, our claim is proved.

Solution by Andy Pepperdine

Consider the arithmetic sequence $y_{n+1} = y_n + k$. By squaring, we have $y_{n+1}^2 = y_n^2 + 2ky_n + k^2$. Add an arbitrary constant a and rearrange to get

$$y_{n+1}^2 + a = (y_n^2 + a) + k^2 + 2ky_n .$$

Setting $x_n = y_n^2 + a$ and $k = 2$, we obtain the desired relation. If y_1, k and a are all integers, then so are all y_n and x_n. In particular, we can set $y_1 = 0$ so that $x_1 = a$.

2. Binary and Dyadic Numerals by Charles Ashbacher, Marion, IA

On page 188 of *The Gödelian Puzzle Book: Puzzles, Paradoxes & Proofs,* Raymond M. Smullyan describes dyadic notation, where all integers greater than zero can be expressed using a string of 1's and 2's. Using the alphabet { 1,2 } all such integers can be expressed in the form

$$2^n d_n + 2^{n-1} d_{n-1} + \cdots + 2^1 d_1 + 2^0 d_0.$$

For example 4, 5 and 6 are 12, 21 and 22 in dyadic form.

Shortly after the description of dyadic notation there is the sentence:

"It is just that for any given number in decimal notation, the digits of the dyadic and binary expressions of the number are different."

a) Determine an infinite family of integers whose binary and dyadic expressions are identical.

b) Prove that no number other than those in the family of part (a) can have identical binary and dyadic expressions.

Reference:

1. Raymond M. Smullyan, *The Godelian Puzzle Book: Puzzles, Paradoxes & Proofs*, Dover Publications, Mineola, New York, 2013. ISBN 978-0-486-49705-1.

Solution by Andy Pepperdine

The binary representation of a number $N > 0$ is:

$$N = \sum_{i=0}^{n} 2^i b_i , b_i \in \{0,1\}, b_n \neq 0.$$

The dyadic representation of N is:

$$N = \sum_{i=0}^{m} 2^i d_i , d_i \in \{1,2\}.$$

These representations will be identical when

$$n = m, \forall i \in \{0,..,n\}, d_i = b_i$$

But, since the d's are selected from $\{1,2\}$ and the b's from $\{0,1\}$ this can happen only when all the digits in each representation are equal to 1.

a) The representation of $2^n - 1$ is all ones, and the same number of ones, in both binary and dyadic forms, and there are an infinite number of these.

b) For all other numbers, the binary form must contain at least one zero bit, and zero is not in the alphabet of dyadic representations. Hence, there are no other identical forms.

3. **Circle Triplet (1)** by Henry Ibstedt, Issy les Moulineaux, France

In Figure 1, a circle of radius a is inscribed in a circle with radius $2a$ and center $(0,0)$, so that it is tangent to the larger circle at the point $(2a,0)$. A circle with center $B(b,d)$ is tangent to these other two circles as well as to the x-axis. Finally, a circle with center $C(c,e)$ is tangent to all three circles mentioned previously. Find the center and radius of this last circle.

Figure 1

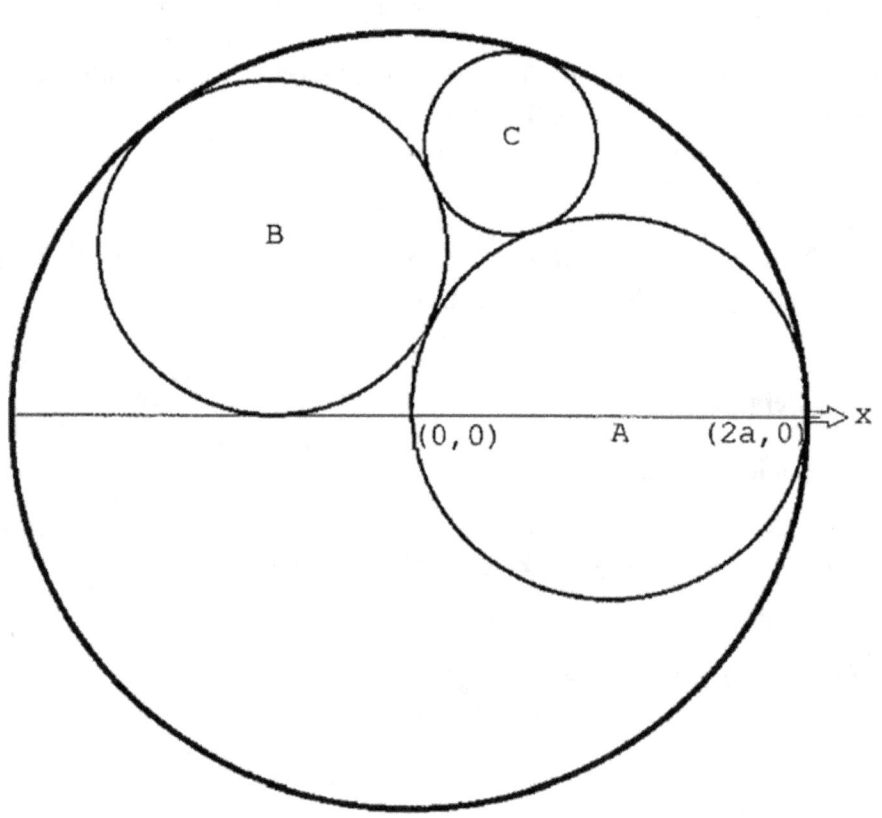

Solution by Andy Pepperdine

The notation $Q(q)$ means the circle with center at Q and radius q.

With the notation in Figure 2, circle $B(b)$ is tangent to the x-axis at H, so angle OHB is a right angle. Let OH be length t. The center of circle $A(a)$ lies on the x-axis and is tangent to the outer circle

From right triangle OHB: $(2a - b)^2 = b^2 + t^2.$ (1)

From right triangle AHB: $(a + b)^2 = b^2 + (a + t)^2.$ (2)

114

Figure 2

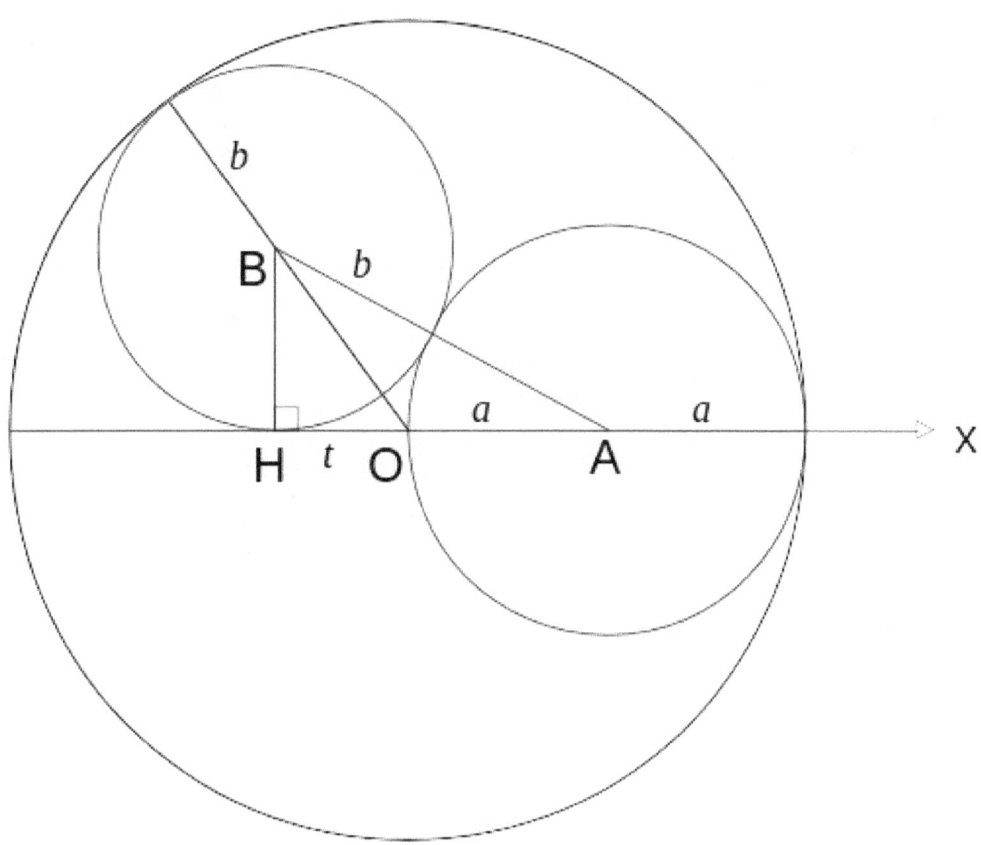

Subtract (1) from (2), and noting that a is not zero, we find that:

$$3a(2b - a) = a(a + 2t)$$

hence,
$$t = 3b - 2a. \qquad (3)$$

Substituting back into (1), and b is not zero:

$$b^2 = 2b(4a - 4b)$$

and so,
$$b = \frac{8}{9}a \qquad (4)$$

From (3) and (4):
$$t = \frac{2}{3}a. \qquad (5)$$

In Figure 3, BM is parallel to the x-axis, and CK is orthogonal to the x-axis. The coordinates of the center of $C(c)$ is at (x,y). We can apply Pythagoras three more times, and do some more algebraic manipulation.

From right triangle OKC: $\qquad (2a - c)^2 = x^2 + y^2 \qquad (6)$

From right triangle AKC: $\qquad (a + c)^2 = (a - x)^2 + y^2 \qquad (7)$

and from right triangle BMC: $\quad (b + c)^2 = (x + t)^2 + (y - b)^2. \qquad (8)$

Figure 3

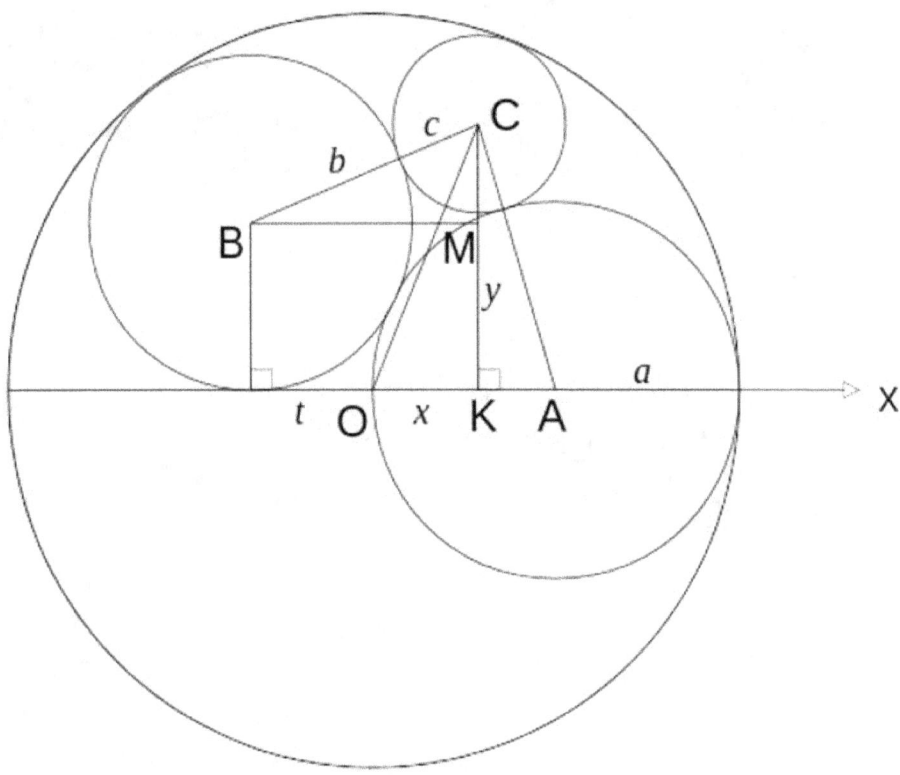

Subtracting (7) from (6) and simplifying:

$$3a(a - 2c) = a(2x - a)$$

which gives: $\qquad\qquad\qquad\qquad x = 2a - 3c.$ $\qquad\qquad$ (9)

Subtracting (8) from (7): $\quad 2by = 2ac - 2bc + 2ax + 2xt + t^2.$

Substituting for b and t from (4) and (5), and simplifying:

$$2y = 8a - 11c.$$ $\qquad\qquad$ (10)

Substituting for x and y in (6) gives a quadratic in c:

$$153c^2 - 208ac + 64a^2 = 0$$

which factors as $\qquad\qquad (9c - 8a)(17c - 8a) = 0.$ $\qquad\qquad$ (11)

From figure 3, it is clear that there will be two solutions, since two circles can be drawn tangent to the outer circle and to both $A(a)$ and $B(b)$. One of them is the reflection of $B(b)$ in the x-axis, by symmetry, which accounts for the first factor in (11). The second factor must apply to the circle $C(c)$ as required.

$$c = \frac{8}{17}a$$

Hence

From (9) and (10) we can now get the co-ordinates of center of C.

Circle C has center at $(10a/17, 24a/17)$ with radius $8a/17$.

Note: We can verify the radius using Descartes' Circle Theorem [1] that the sum of the squares of the reciprocals of the radii (the bends, in Soddy's terminology [2]) of four mutually tangent circles is half the square of their sum, bearing in mind that circles that are touched internally are given negative radii. In this case, the bends are -1/2, 1, 9/8 and 17/8. Multiplying through by 8, we see that

$$(-4)^2 + 8^2 + 9^2 + 17^2 = 450 = \frac{1}{2}(-4 + 8 + 9 + 17)^2$$

References

[1] Coxeter, H.S.M, *Introduction to Geometry*, sec 1.57, p 14, Wiley, 1969

[2] Soddy, F, *The Kiss Precise*, Nature, 137, p 1021, 1936

There are several on-line proofs of Descartes Circle Theorem, for example:
http://euler.genepeer.com/from-herons-formula-to-descartes-circle-theorem/

4. Circle Triplet (2) by Henry Ibstedt, Issy les Moulineaux, France

In Figure 4, a circle with center $A(a,0)$ is inscribed in a circle with radius $2a$ and center $(0,0)$, so that it is tangent to the larger circle at the point $(2a,0)$. A circle with center $B(b,d)$ is tangent to these other two circles as well as to the x-axis. Finally, a circle with center C is tangent to the x-axis and to the circles with centers A and B. Find the center and radius of this last circle.

Solution by Andy Pepperdine

This solution uses a different technique from that used in question 3.

The notation $Q(q)$ means the circle with center at Q and radius q, and $Q'(q')$ is the inverse of Q in the circle of inversion $I(i)$. With the notation in Figure 5, we use a circle of inversion $I(i)$ with center at $(2a, 0)$ and radius $2a = i$. Circle $A(a)$ inverts to line A' since $A(a)$ passes through the center of inversion I and is tangent to $I(i)$ at O. Circle $O(o)$ inverts to line O', since it passes through the center of inversion, and by symmetry (the radii of $O(o)$ and $I(i)$. are equal), O' passes through A as well as the points of intersection of $O(o)$ and $I(i)$. The x-axis inverts to itself.

Circle $B(b)$ is tangent to each of the x-axis, and circles $O(o)$ and $A(a)$. Hence its inverse $B'(b')$ touches the x-axis, and lines A' and O'. So its diameter is a.

Let $B(b)$ touch the x-axis at H, then $B'(b')$ touches the axis at H', and hence

$$IH' = \frac{3a}{2} \quad \text{and} \quad B'H' = b' = \frac{a}{2}.$$

Figure 4

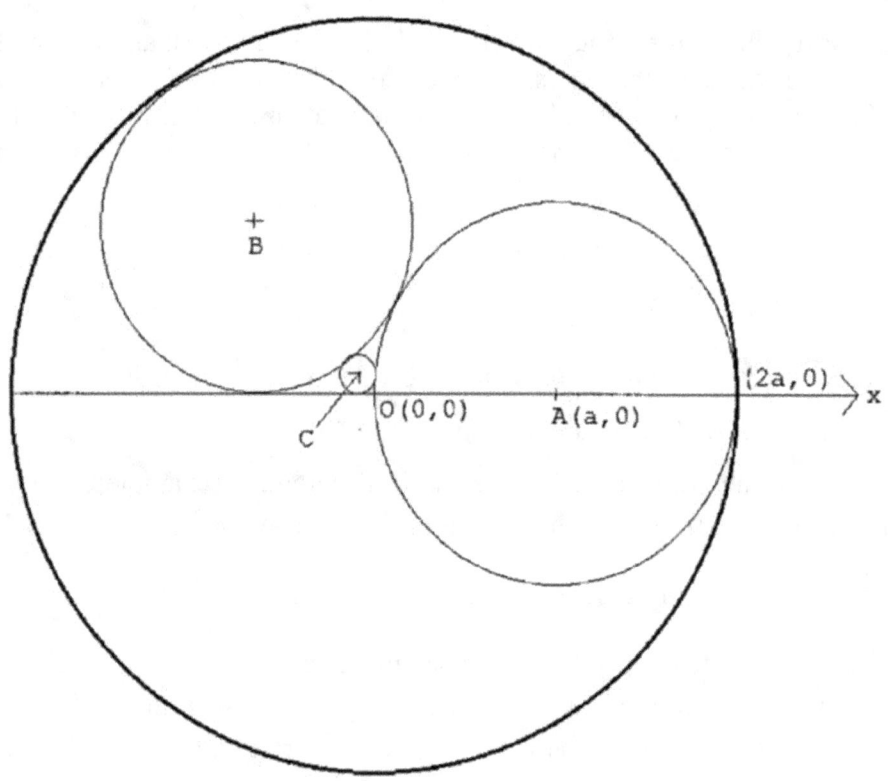

Figure 5

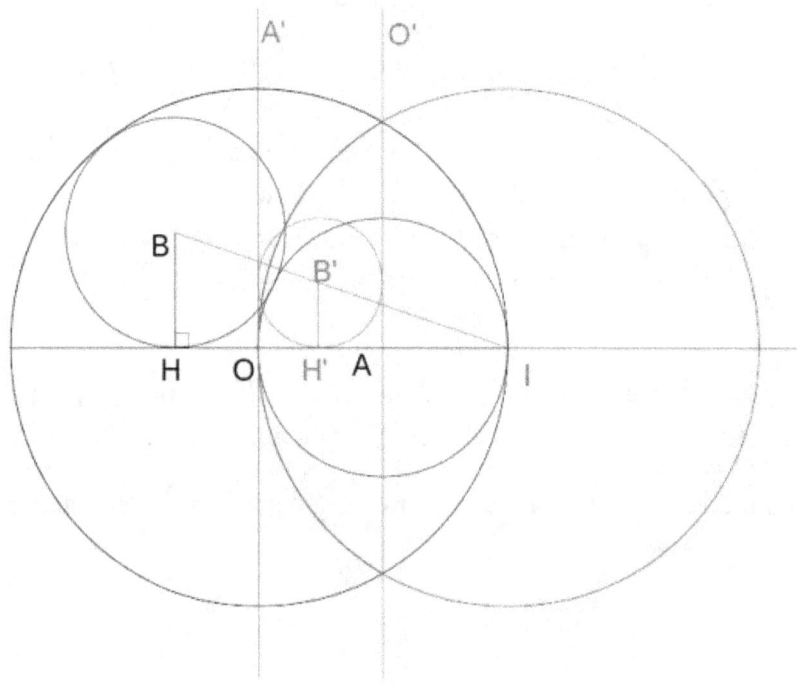

By the properties of inversion, $IH \cdot IH' = (2a)^2 = 4a^2$.

Hence $IH = \frac{8a}{3}$.

Now I, B and B' are collinear. So by similar triangles,
$BH/BH' = IH/IH'$, which gives $b = BH = \frac{8a}{9}$.

The circle $C'(c')$ is tangent to the x-axis, and circles $A(a)$ and $B(b)$. So its inverse touches the x-axis, line A' and circle $B'(b')$ and hence it nestles in the corner at the origin of the coordinate system.

From a simple application of Pythagoras we can get

$$(b' + c')^2 = 2(b' - c')^2 .$$

But we want the smaller solution to this, so we deduce that $c' = \left(3 - 2\sqrt{2}\right)\frac{a}{2}$.

To keep the diagram clean, the circle $C(c)$ and its inverse $C'(c')$ will not be drawn, but let them touch the x-axis at N and N' respectively. We can now proceed to find the location of C in the same way we did for B.

$IN' = 2a - c' = \left(1 + 2\sqrt{2}\right)\frac{a}{2}$ and $IN \cdot IN' = 4a^2$ and so

$IN = 8\left(2\sqrt{2} - 1\right)\frac{a}{7}$.

Then $IN/IN' = CN/C'N'$ giving $c = CN = \frac{8a}{49}\left(43 - 30\sqrt{2}\right)$.

Thus the center of circle $C(c)$ has coordinates

$$\left(\frac{2a}{7}\left(11 - 8\sqrt{2}\right), \frac{8a}{49}\left(43 - 30\sqrt{2}\right)\right) .$$

5. Wall Scraper by Hubert Hagadorn, Menlo Park, CA

A semicircle of diameter 2 is able to move along a path of unit width having a sharp right angle turn, sliding, rotating and then sliding again. What is the shape of largest area that is able to travel along this path and negotiate the turn?

Solution by the Proposer

Not known. The area of the semicircle is about 1.571, but larger areas are also possible. Figure 6 shows the right side of a symmetrical shape for a region having an area of 2.218. Areas were in agreement to four decimal places for calculations based on ellipse formulas, and those based on calculation at about 4170 discrete points along the perimeter.

Figure 6

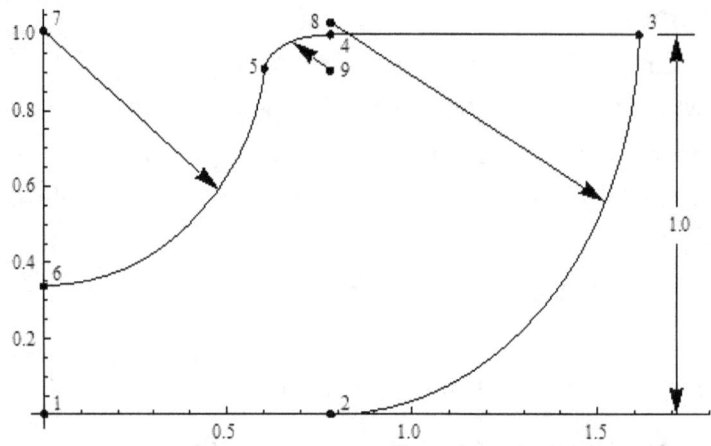

The region is defined by two lines and arcs of three ellipses. Horizontal lines are from point 1 to 2, and from point 3 to 4. The centers for ellipses 1, 2 and 3, are at points 7, 8, and 9, respectively. The slopes of ellipses 1, 2, and 3 are zero at points 6, 2, and 8, respectively. Points 2, 4, 8, and 9 have the same x-coordinate. Ellipses 1 and 3 are tangent at point 5.

In Figure 7 the region is shown at four positions as it rotates from zero degrees to 45 degrees in 15 degree increments.

Table 1 gives the semi-axes of the three ellipses, where a and b refer to their axes aligned with the x and y axes, respectively. Table 2 gives the coordinates of the nine points. The seven values with asterisks are those parameters optimized. Other values excluding 0 and 1 are derived values.

Table 1

Ellipse	a	b
1	*0.612140	*0.670108
2	*0.830403	*1.031406
3	0.177115	*0.094859

Figure 7

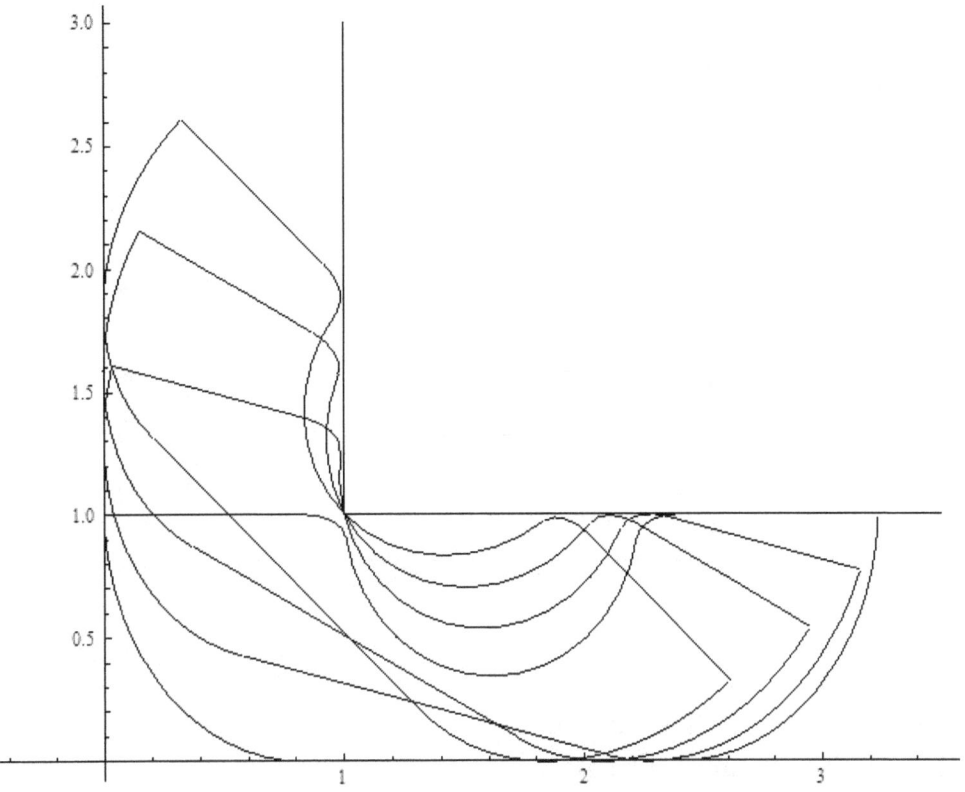

6. **Three-Section Square Inside an Equilateral Triangle** by Hubert Hagadorn, Menlo Park, CA

A square is cut in three sections so as to fit inside an equilateral triangle with maximum area coverage. The three pieces may not overlap. How should the square be cut and what is the percentage of area coverage? Henry Dudeney has shown that the answer is 100% if the square is allowed to be cut into four pieces.

Table 2

Point	x	y
1	0	0
2	0.78233	0
3	1.61235	1
4	0.78233	1
5	0.605669	0.911926
6	0	0.339000
7	0	*1.009108
8	0.78233	1.031406
9	0.78233	*0.905141

Solution by the Proposer

Two possible solutions are given, as illustrated in the two pairs of figures. Both give the same ratio of triangle side length to square side length, and hence give the same coverage. Other solutions giving the same results are believed to exist.

Referring to the first figure pair,

$\alpha = 30$ degrees (also valid for the second figure pair),

$a_s = (3 - \sqrt{3})a_t / 2 \cong 0.634 a_t$,

$b = a_s / 2$.

The square area is $A_s = a_s^2 = 3(2 - \sqrt{3})a_t^2 / 2$, and triangle area is $A_t = \sqrt{3}a_t^2 / 4$.

Taking the ratio of the square area to the triangle area, the percent coverage is

$200(2\sqrt{3} - 3) \cong 92.82 \%$.

First solution

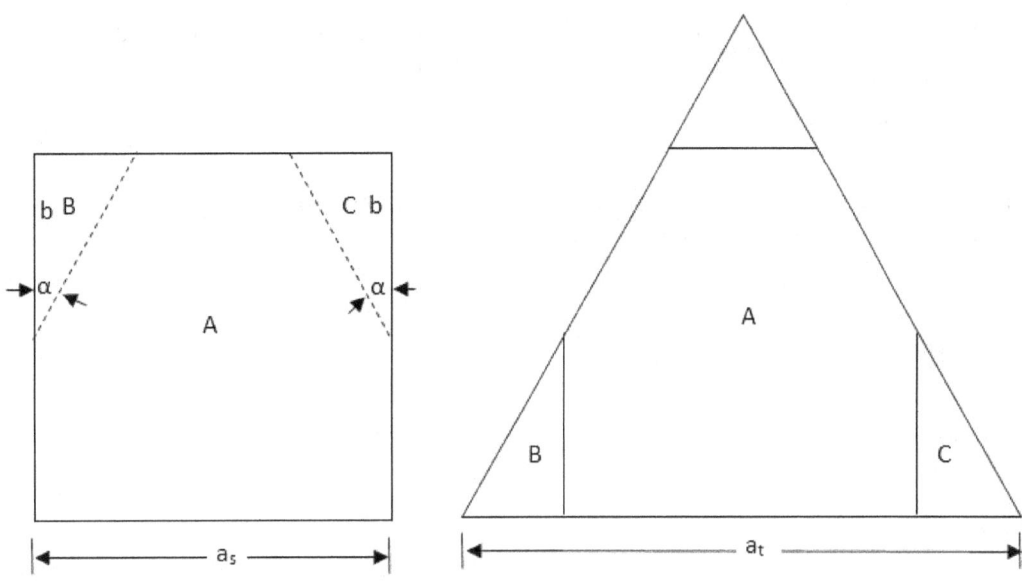

Second solution

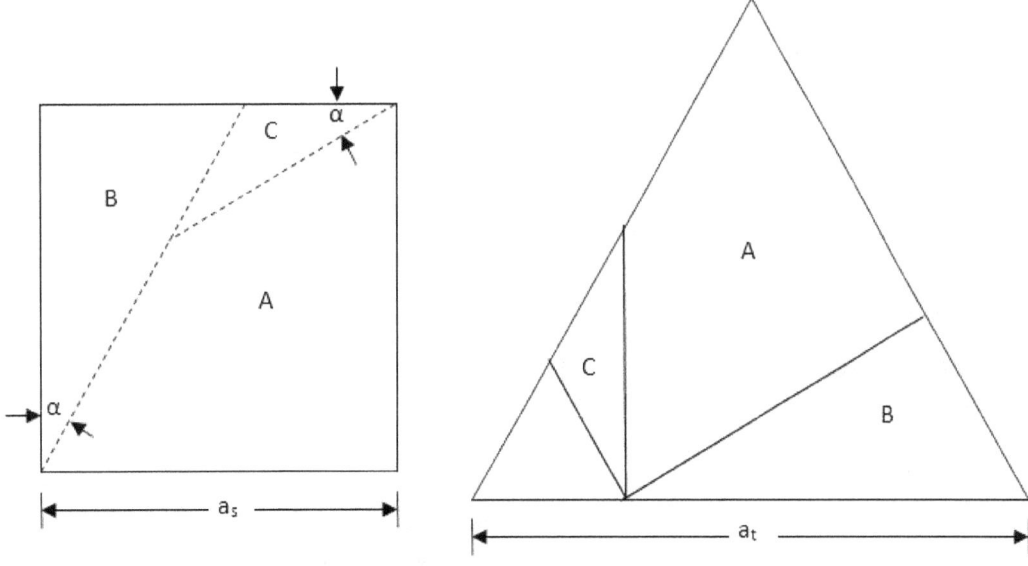

123

7. Three-Section Equilateral Triangle Inside a Square by Hubert Hagadorn, Menlo Park, CA

An equilateral triangle is cut in three sections so as to fit inside a square with maximum area coverage. The three pieces may not overlap. How should the triangle be cut and what is the percentage of area coverage?

Solution by the Proposer and by Brian Barwell (Independently)
The section cuts are indicated in Figure 8, where

$$a_t = 7a_s(1 - \sqrt{3}/3)/2 \cong 1.479a_s,$$
$$b = (5\sqrt{3} - 7)a_s/2 \cong 0.830a_s.$$

The triangle area is $A_t = 49a_s^2(2\sqrt{3} - 3)/24,$

and square area is $A_s = a_s^2.$

Taking the ratio of the triangle area to the square area, the percent coverage is

$$\frac{4900(2\sqrt{3}-3)}{24} = 94.75\ \%\ .$$

Figure 8

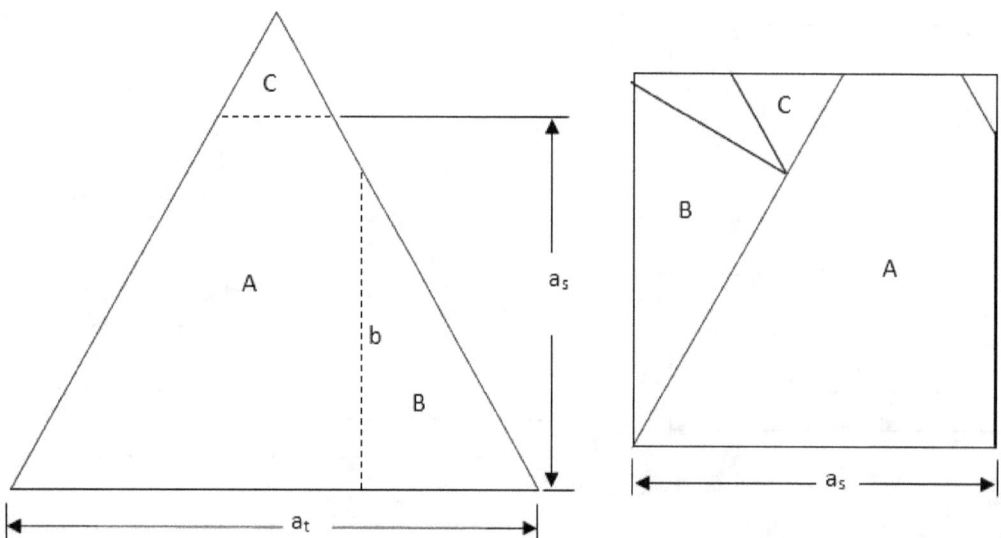

8. Scrambled Math Words by Hubert Hagadorn, Menlo Park, CA

Unscramble the letters in *laimgotlhcyrial* to obtain two words, each using all fifteen letters.

Solution by Several Readers

The words are ALGORITHMICALLY and LOGARITHMICALLY.

9. Catch Twenty-Two by Robert Wainwright, New Rochelle, NY

Four students (A, B, C and D) enter a three-day math competition and each is assigned two integers to be used throughout. A is assigned 2 and 9, B gets 3 and 8, C gets 4 and 7 while D has 5 and 6. Each day they are paired in teams of two and the pairings are never repeated. Each day the team's task is to use their four digits to form expressions which equal 22. They may concatenate digits as well as using the operators $+$, $-$, $\times$, $\div$ and $\sqrt{}$. Each expression must use all four of their digits and operators may be used more than once in any expression. The radical symbol may not have an index, so it indicates a square root only.

a. Determine the number of possible solutions for each team.
b. Determine the total number of possible solutions by each student over the three days.

Solution by Hubert Hagadorn

The digit pairs for A, B, C, D, are
A: 2, 9
B: 3, 8
C: 4, 7
D: 5, 6

There are six possible pairings
AB: 2,3,8,9
AC: 2,4,7,9
AD: 2,5,6,9
BC: 3,4,7,8
BD: 3,5,6,8
CD: 4,5,6,7

Hence two of these pairs may be selected for each of the three days, no pairings repeated. One may note that the sum of all pairings is 22, all numbers are used, and so this is one solution for each pairing (of pairs). Let the number of solutions for part (a) corresponding to AB, AC, AD, BC, BD, CD be n1, n2, n3, n4, n5, n6, respectively. Then the number of solutions for each student, part (b), is A: n1 + n2 + n3, B: n1 + n4 + n5, C: n2 + n4 + n6, and D: n3 + n5 + n6, For the first solution considered it is assumed that multiple instances of each digit is allowed. Let the digits for a given pairing be given as d1, d2, d3, and d4. Also let a concatenation of n digits of d1 be given as $d1_n$, and similarly for d2, d3, d4. Taking advantage of the fact that the sum of the four digits equals 22, a formula for other sums that has the property that no number is ever repeated is

$$\frac{d1_{2n}}{d1_{2n(2n+1)}}(d1_{2n+1} + d2_{2n+1} + d3_{2n+1} + d4_{2n+1}), \ n = 1, 2, 3, \dots.$$

In this case the solutions for parts (a) and (b) are infinite.

The second solution set assumes that a given digit is used only once. Possibly there are additional solutions than those given.

For AB:

1. $2 + 3 + 8 + 9$
2. $9 - 3 + 2 * 8$
3. $9 * 8/3 - 2$
4. $3 + \sqrt{9} + 2 * 8$
5. $2(8 + 9/3)$
6. $3(8 - 2/\sqrt{9})$
7. $\sqrt{9}(8 - 2/3)$
8. $23 + 8 - 9$
9. $28 + 3 - 9$
10. $28 - 3 - \sqrt{9}$
11. $\sqrt{9} + 38/2$
12. $23 - \sqrt{9} - 8$

For AC:

1. $2 + 4 + 7 + 9$
2. $4 * 9 - 2 * 7$
3. $4 * 7 - 2 * \sqrt{9}$
4. $7 * \sqrt{9} + 2/\sqrt{4}$
5. $7 * \sqrt{9} + \sqrt{4}/2$
6. $2 + \sqrt{4}(7 + \sqrt{9})$
7. $\sqrt{4} + 2(7 + \sqrt{9})$
8. $2(7 * \sqrt{4} - \sqrt{9})$
9. $2(9 * \sqrt{4} - 7)$
10. $4(7 - \sqrt{9}/2)$
11. $4(9 - 7/2)$
12. $\sqrt{4}(2 * 7 - \sqrt{9})$
13. $\sqrt{4}(2 * 9 - 7)$
14. $24 + 7 - 9$
15. $27 + 4 - 9$
16. $27 - \sqrt{4} - \sqrt{9}$
17. $72/\sqrt{9} - \sqrt{4}$
18. $24 - \sqrt{7 - \sqrt{9}}$
19. $(47 - \sqrt{9})/2$
20. $49 - 27$
21. $94 - 72$

For AD:

1. $2 + 5 + 6 + 9$
2. $6 * 9/2 - 5$
3. $2(5 + 9) - 6$
4. $6 + 2(5 + \sqrt{9})$
5. $6(9 - 5) - 2$
6. $6(2 + 5/\sqrt{9})$
7. $25 + \sqrt{9} - 6$
8. $25 - 9 + 6$
9. $25 - \sqrt{6 + \sqrt{9}}$
10. $26 - 9 + 5$
11. $\sqrt{625} - \sqrt{9}$

For BC:

1. $3 + 4 + 7 + 8$
2. $4 * 8 - 3 - 7$
3. $7 * 8 - 34$
4. $78/3 - 4$
5. $(74 - 8)/3$
6. $\sqrt{487 - 3}$

For BD:

1. $3 + 5 + 6 + 8$
2. $5 * 8 - 3 * 6$
3. $6 + 8(5 - 3)$
4. $5 * \sqrt{36} - 8$
5. $58 - 36$
6. $85 - 63$

For CD:

1. $4 + 5 + 6 + 7$
2. $5 - 7 + 4 * 6$
3. $7 + 5 * 6 / \sqrt{4}$
4. $6 * 7 - 4 * 5$
5. $67 - 45$
6. $76 - 54$
7. $\sqrt{576} - \sqrt{4}$

a. The number of solutions for AB, AC, AD, BC, BD, CD are 12, 21, 11, 6, 6, 7, respectively.
b. The number of solutions for students A, B, C, D, are 44, 24, 34, 24, respectively.
Only one permutation of a sum or product is given (There are 24 permutations of the first solution for each team.). Also, only one permutation of the form a*b/c, a/b*c, or a/b/c is given.

Editor's Commentary

All solvers, including the proposer, made assumptions about what constituted a "different" expression. **Hagadorn** was the only one who considered the case of repeated use of a digit. **Andy Pepperdine** assumed that each digit must be used exactly once in an expression and clearly specified his meaning of different expressions. His counts were similar to Hagadorn's.

10. Cevian Construction by Kostantinos Myrianthis, Athens, Greece

Let ABC be a triangle and let AA' be a cevian inside the triangle. Construct cevians BB' and CC' such that they intersect at the point P on AA' and also have the property that $AB' = AC'$.

References

1. R.L. Goodstein, E.J. Primpose, *Axiomatic Projective Geometry,* University College Leicester, 1953

2. J.G. Semple, G.T. Kneebone, *Algebraic Projective Geometry,* Oxford at the Clarendon Press, 1979

Solution by the Proposer

Let BB_0 and CC_0 be two cevians which intersect inside the triangle at P_0 on AA', as seen in figure 9. We extend C_0B_0 and let R be the intersection of C_0B_0 and BC. The points B, A', C, R form a harmonic set (see [1],[2]), so :

$$(R,A';B,C) = -1$$

Figure 9

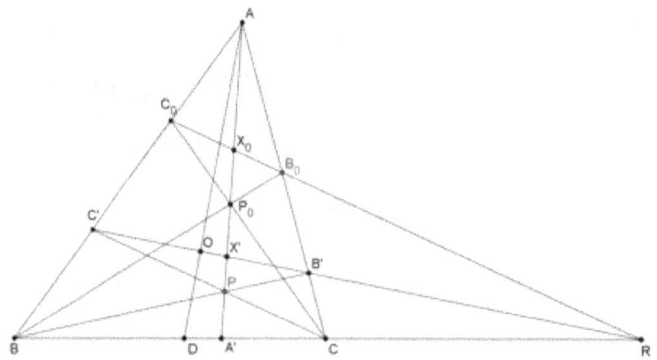

More precisely, A' is the harmonic conjugate of R with respect to B and C. Let X_0 be the intersection point of AA' and C_0B_0. The points C_0, X_0, B_0, R form also a harmonic set, so :

$$(R,X_0;C_0,B_0) = -1$$

Let AD be the bisector of the angle $\angle BAC$ and let RO be the perpendicular line segment to AD at point O from point R, which intersects the line segments AB at C', AC at B' and AA' at X'. The points C', X', B', R form a harmonic set, so :

$$(R,X';C',B') = -1$$

As a corollary of the above, the cevians BB' and CC' intersect at point P on AA'. Finally we have $AC'=AB'$ (because RO is perpendicular to the bisector AD).

References:

1. R.L. Goodstein, E.J. Primpose, Axiomatic Projective Geometry, University College Leicester, 1953

2. J.G. Semple and G.T. Kneebone, Algebraic Projective Geometry, Oxford at the Clarendon Press, 1979

11. Negative Bases by Brian Barwell, Hampton, Middlesex, UK

Integers are often expressed in bases other than 10 such as octal or hexadecimal but integers expressed in negative bases are a lot less familiar though they work in the same way, for example, using subscript notation to indicate bases, we find in base -8

$$(325)_{-8} = 3 \times (-8)^2 + 2 \times (-8) + 5 = 192 - 16 + 5 = 181$$

Negative bases have some unfamiliar properties. An integer with an even number of digits represents a negative number, e.g.

$$(1234)_{-11} = 1 \times (-11)^3 + 2 \times (-11)^2 + 3 \times (-11) + 4 = -1118$$

Another property is that every integer has two representations in a given negative base, one with a minus sign and one without. For example,

$$(345)_{-10} = 300 - 40 + 5 = 265$$

$$(1875)_{-10} = -1000 + 800 - 70 + 5 = -265$$

So in base -10, both 345 and -1875 represent the decimal number 265.

Discover another interesting property of negative bases by expressing the decimal number 147765318 in base -11.

Solution by the Proposer

The decimal number 147765318 when converted into base -11 becomes 147765318 showing that if negative bases are considered, it is possible for an integer to have the same digital representation in two different bases.

Editor's Commentary

I thank **Doug Robertson** of the University of Colorado for the following communication regarding standard practice with negative bases:

I've enjoyed your book on recreational mathematics, but I have an objection to the discussion about negative radix arithmetic on page 75. Mr. Barwell uses a negative sign to argue that $-1875 = 265$. This may be correct in some technical sense, but the whole point of negative radices is to avoid the use of negative signs altogether. They are not needed, and only cause confusion with no offsetting benefit. You might as well argue that there are two representations of 265 in decimal arithmetic because $-(-265) = 265$. Donald Knuth, in The Art of Computer Programming, v. 2, pp. 188-190, has a detailed discussion of negative radix arithmetic and the relation to such things as two's complement arithmetic, which is actually used in computer systems and eliminates the need for a sign bit as well. Knuth also has a discussion of imaginary radices, $2i$ and $i-1$.

12. Pythagorean Dissection by Brian Barwell, Hampton, Middlesex, UK

Find a four-piece dissection of a 7×7 square and a 24×24 square such that the pieces can be re-assembled to form a 25×25 square. Pieces may be turned over but the cuts may only be made along boundaries of the unit squares which make up the larger squares.

Solution by Hubert Hagadorn

Use three parallel cuts to divide the 7×7 square into parts A, B and C, each with dimensions 7×2, and part D with dimensions 7×1 .Dissect the 24×24 square as shown in figure 10.

Figure 10

E is has dimensions 23×24 while F and G have dimensions 3×1, and H dimensions 18×1. Assemble the 25×25 as shown in figure 11.

Figure 11

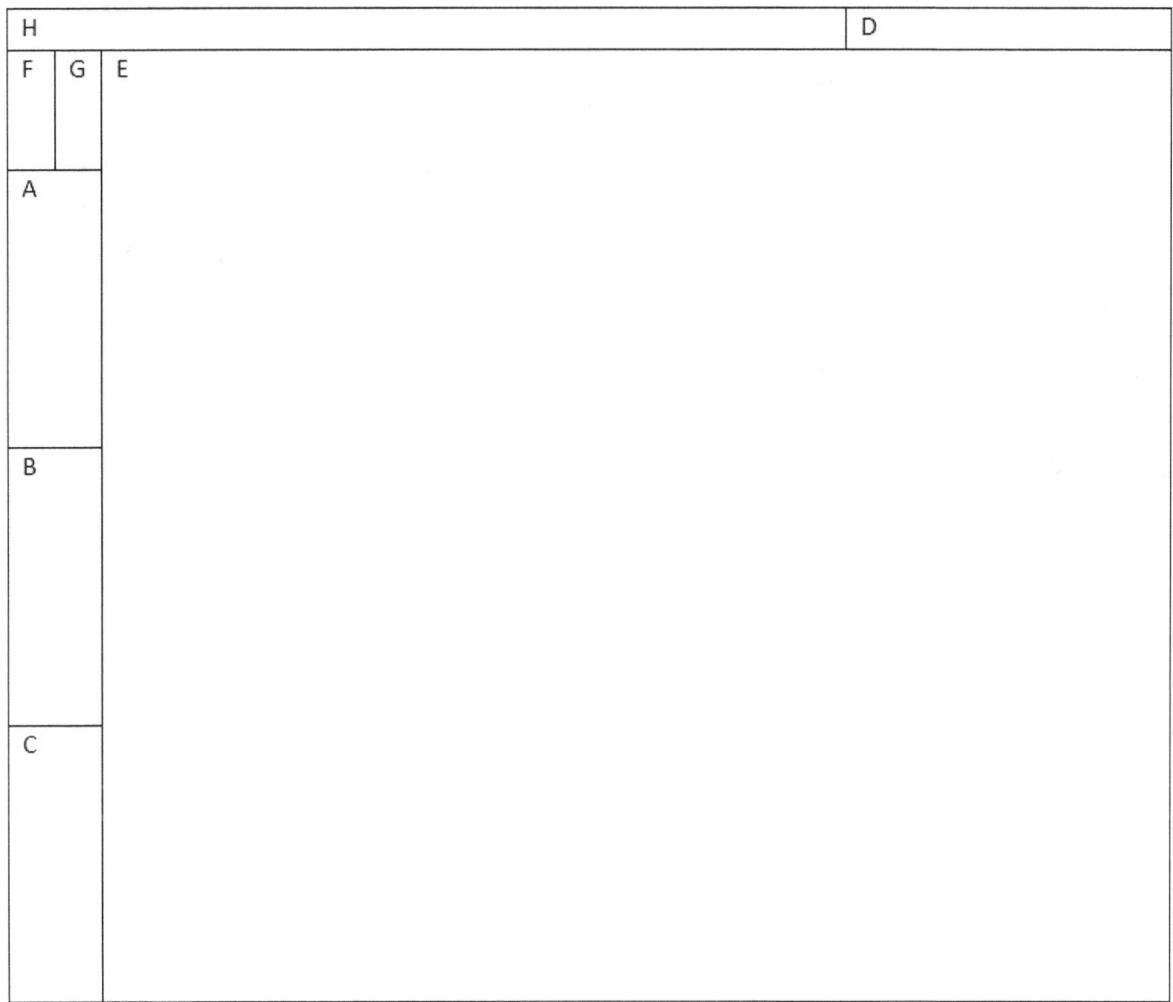

Editor's Commentary

It is clear from the submissions which I received that the intent of the proposer was unclear to our readers. His solution cuts the two 7×7 and 24×24 squares into a <u>total</u> of four pieces (which are not all rectangles). Readers are invited to search for solutions to the intended problem.

Proposers And Solvers List For Problems Which Appeared In Topics In Recreational Mathematics Volume 1

P	S			1	2	3	4	5	6	7	8	9	10	11	12
■		Charles Ashbacher	Marion, IA		P										
■	■	Brian Barwell	Hampton, Middlesex, UK	S	S	S	S			S	S			P	P
		Michael P. Cohen	Washington DC	S											
■	■	Hubert Hagadorn	Menlo Park, CA				P	P	P	P	P	S			S
■		Henry Ibstedt	Issy les Moulineaux, France	P	S	P	P		S	S		S		S	S
	■	Kathleen Lewis	Brikama, Gambia	S											
■	■	Kostantinos Myrianthis	Athens, Greece										P		
	■	Andy Pepperdine	Bath, UK	S	S	S	S				S	S			
■		Robert Wainwright	New Rochelle, NY									P			
■	■	Proposer/Solver													

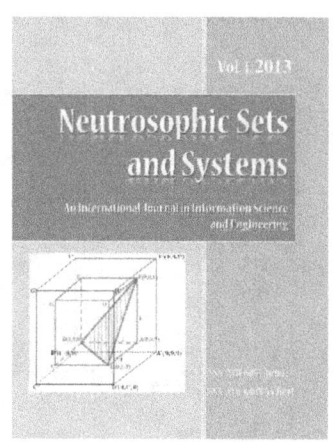

Neutrosophic Sets and Systems has been created for publications on advanced studies in neutrosophy, neutrosophic set, neutrosophic logic, neutrosophic probability, neutrosophic statistics that started in 1995 and their applications in any field, such as the neutrosophic structures developed in algebra, geometry, topology, etc.

The submitted papers should be professional, in good English, containing a brief review of a problem and obtained results. <u>Neutrosophy</u> is a new branch of philosophy that studies the origin, nature, and scope of neutralities, as well as their interactions with different ideational spectra.

This theory considers every notion or idea <A> together with its opposite or negation <antiA> and with their spectrum of neutralities <neutA> in between them (i.e. notions or ideas supporting neither <A> nor <antiA>). The <neutA> and <antiA> ideas together are referred to as <nonA>.

Neutrosophic Set and Logic are generalizations of the fuzzy set and respectively fuzzy logic (especially of intuitionistic fuzzy set and respectively intuitionistic fuzzy logic). In neutrosophic logic a proposition has a degree of truth (T), a degree of indeter

minacy (I), and a degree of falsity (F), where T, I, F are standard or non-standard subsets of $]^-0, 1^+[$.

Neutrosophic Probability is a generalization of the classical probability and imprecise probability.

Neutrosophic Statistics is a generalization of the classical statistics.

What distinguishes the neutrosophics from other fields is the <neutA>, which means neither <A> nor <antiA>.

<neutA>, which of course depends on <A>, can be indeterminacy, neutrality, tie game, unknown, contradiction, ignorance, imprecision, etc.

All submissions should be designed in MS Word format using our template file:

http://fs.gallup.unm.edu/NSS/NSS-paper-template.doc

A variety of scientific books in many languages can be downloaded freely from the Digital Library of Science:

http://fs.gallup.unm.edu/eBooks-otherformats.htm

To submit a paper, mail the file to the Editor-in-Chief. To order printed issues, contact the Editor-in-Chief. This journal is non-commercial, academic edition. It is printed from private donations.

Information about the neutrosophics you get from the UNM website:

http://fs.gallup.unm.edu/neutrosophy.htm

The home page of the journal is accessed on

http://fs.gallup.unm.edu/NSS

BOOKS IN RECREATIONAL MATHEMATICS BY CHARLES ASHBACHER AND ASSOCIATES

Topics in Recreational Mathematics 1/2015 ISBN 978-1507603215

Topics in Recreational Mathematics 2/2015 ISBN 978-1508617099

Topics in Recreational Mathematics 3/2015 ISBN 978-1511641005

Topics in Recreational Mathematics 4/2015 ISBN 978-1514317518

Alphametics as Expressed in Recreational Mathematics Magazine ISBN 978-1508538134

Ten Year Cumulative Index to the Journal of Recreational Mathematics, edited by Joseph S. Madachy and Charles Ashbacher ISBN 978-1508936800

Alphametics Expressing Thoughts From the Star Trek Original Series ISBN 978-1512152784

Mathematical Cartoons ISBN 978-1514207130

Solved Problems in Statistical Inference ISBN 978-1515215622

Associates

Artist Catie Ribble

Editor Rachel Pollari

Editor Jennifer Corrigan

Artist Jenna Richardson